AutoCAD 2012 with AutoLISP

An introductory guide

BASUDEB BHATTA

ISBN: 1463625804
ISBN-13: 978-1463625801

Dedicated to my daughter *Bagmi*

Preface

This book has been designed for the beginners of CAD users. This comprehensive book offers a hands-on, activity-based approach to the use of AutoCAD—complete with techniques, tips, shortcuts, and insights designed to increase the efficiency. The book not only covers 2D drawing, but also focuses on isometric and 3D drawings in addition to AutoLISP programming. AutoLISP is a dialect of Lisp programming language built specifically for use with the full version of AutoCAD and its derivatives. Books on AutoLISP are very limited. This book provides the basic understanding of AutoLISP programming with several examples.

Topics and tasks have been carefully grouped to guide students logically through the AutoCAD command set, with the level of difficulty increasing steadily as skills are acquired through experience and practice. Straightforward explanations focus on what is relevant to actual drawing procedures. Several illustrations show exactly what to expect on the computer screen. This book also features several exercises that help the students to assess their skill and comprehension about the subject matter studied in this book.

The primary purpose of this book is to be a learning resource for college and university students, as well as for individuals now in the industry who require indoctrination in AutoCAD and AutoLISP.

Acknowledgments

I am very much thankful to Prof Rana Dattagupta, Director, Computer Aided Design Centre, Computer Science and Engineering Department, Jadavpur University, Kolkata for extending necessary facilities to write this book. I am thankful to my colleagues, especially Mr Biswajit Giri, Mr Chiranjib Karmakar, Mr Subrata Das, Mr Santanu Glosal, and Mr Uday Kumar De. Without their help and cooperation writing of this book was never possible.

I would like to express my gratitude to my parents who have been a perennial source of inspiration and hope for me. I also want to thank my wife Chandrani, for her understanding and full support, while I worked on this book. My little daughter, Bagmi, deserves a pat for bearing with me during this rigorous exercise.

B. Bhatta

Contents

Chapter 3 Display Techniques and Dimensioning

Chapter 4 Advanced Techniques

Chapter 5 Isometric and 3D Drawings

1 Primitives

Introduction

CAD stands for Computer Aided Design or Computer Aided Drawing. AutoCAD®, developed by Autodesk Inc (www.autodesk.com), is the most popular PC-based computer aided design and drafting system available in the market. First released in December 1982, AutoCAD was one of the first CAD programs to run on personal computers. Currently, more than 5 million people around the world are using AutoCAD to generate two-dimensional (2D) and three-dimensional (3D) drawings. It is very user-friendly, and most popular for any type of engineering drawing. To fully appreciate its benefits, one should think of AutoCAD as not just a fancy drafting tool, but means the *modeling a design* on the computer. AutoCAD offers a higher level of speed and accuracy; and it is easy to use. It provides drawing accuracy of 16 decimal places.

AutoCAD's native file format is DWG, and its interchange file format is DXF. These two formats have become de facto standards for CAD data interoperability. AutoCAD in recent years has included support for DWF, a format developed and promoted by Autodesk for publishing CAD data.

Generally, each year a new version of the software is released. The latest version of AutoCAD is 18.2, commonly known as AutoCAD 2012. It was released on March 2011.

System Requirement for AutoCAD 2012

To run AutoCAD 2012 on Windows®, the following minimum software and hardware are required (Autodesk-recommended for good performance):

Operating System (any one)	Microsoft Windows XP, Service Pack 3
	Microsoft Windows Vista or Windows 7
Browser	Microsoft Internet Explorer 7.0 or later
Processor	Pentium-IV or later, 1.6 GHz or higher
RAM	1 GB (minimum), 2 GB (recommended)
Video	1024 × 768 with True Color
Hard Disk	Installation 2 GB
Pointing Device	Mouse, trackball or other device

A First Look at AutoCAD

Now let us take a tour of AutoCAD. In this section you will get a chance to familiarize yourself with the AutoCAD screen and to communicate with AutoCAD.

Starting AutoCAD

Launch AutoCAD by clicking on **AutoCAD 2012** icon on Windows desktop. It can also be launched by clicking on Windows **Start** menu and then pointing to **Programs** → **Autodesk** → **AutoCAD 2012**. Now the AutoCAD drawing editor (Figure 1.1) will appear on the screen.

The Drawing Editor

Once the AutoCAD is launched, the initial screen displays is called **drawing editor**. It contains the following main items: (1) Menu bar (pull-down menus), (2) status bar, (3) drawing window, (4) command prompt window, and (5) tool bars.

The AutoCAD **menu bar** contains a list of commands and options. Commands are issued by pulling down a menu by clicking on it and selecting the command by the mouse. After the initial installation one may not see this menu bar. AutoCAD 2012 provides four different workspaces: (1) Drafting & Annotation, (2) 3D Basics, (2) 3D Modeling, and (3) AutoCAD Classic. **AutoCAD Classic** may be preferred by the beginners and those who are habituated with the appearance of earlier versions. Preferred workspace can be chosen using **Workspace Switching** tool (refer Figure 1.1). Initially, the drawing area is filled with equally spaced grid lines. These lines can be removed by clicking on **Grid** tool (refer Figure 1.1).

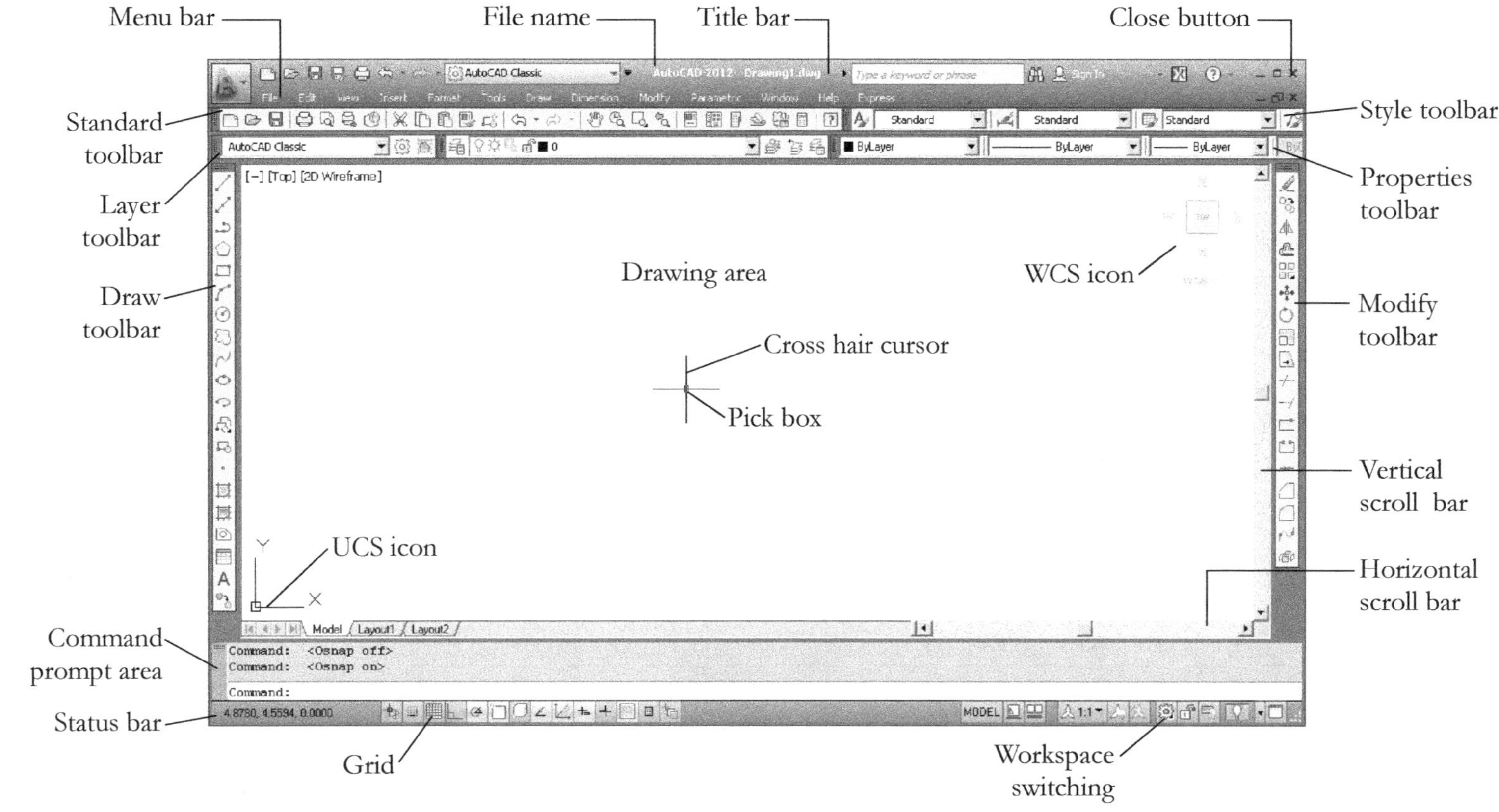

Figure 1.1 The drawing editor of AutoCAD 2012

The **status bar** displays the cursor coordinates and the status such as grid, snap mode (discussed later), etc. Mode tools are always visible in the status bar as selectable buttons. One can turn them ON or OFF by clicking on them. The AutoCAD **drawing window** is used to draw and display the drawings. The **command prompt** is used for entering commands by keyboard. **Tool bar** contains a set of icons representing frequently used commands. Moving the mouse pointer over an icon displays a tooltip with the name of the command. Simultaneously a description of the command's function is also displayed, if the mouse pointer is placed on the tool for three seconds. To invoke a command, one has to click on the icon.

Coordinate System

In the AutoCAD's drawing editor we have to work with the concept of coordinate system. The default limits of work is 12, 9. The lower-left corner of your drawing area is 0,0 and upper-right corner is 12, 9. That means, space provided along *X*-direction is 12 unit and along *Y*-direction is 9 unit (one can change the limits; discussed later).

In AutoCAD, coordinate system is based on Cartesian and Polar coordinate systems. These two coordinate systems can be used in absolute sense as well as relative sense.

> **Note:** 'Default' means which is initially supplied or assumed by the software. In AutoCAD, the default option or value rests in an angular bracket like <default>.

Cartesian Coordinate System

A Cartesian coordinate system specifies each point uniquely in a plane by a pair of numerical coordinates, which are the signed distances from the point to two fixed perpendicular directed lines (*X*-axis and *Y*-axis), measured in the same unit of length. The input form of this system is *X*,*Y*. Figure 1.2 shows the system. There are four quadrants. In the first quadrant *X* and *Y* both are positive; in the second quadrant *X* is negative and *Y* is positive; in the third quadrant *X* and *Y* both are negative; and in the 4th quadrant *X* is positive and *Y* is negative. The crossing point of *X*-axis and *Y*-axis is called *origin*. The coordinate of the origin is 0,0. A third axis, Z (perpendicular to the *X* and *Y*), is also considered to represent 3D objects.

If we want to represent a point in this system, we have to give the *X* and *Y* coordinates of specified point. For example, a point is at 3,-5; that means, the point is situated in fourth quadrant, and it is 3 units from the origin along +*X* direction and 5 units from origin along -*Y* direction.

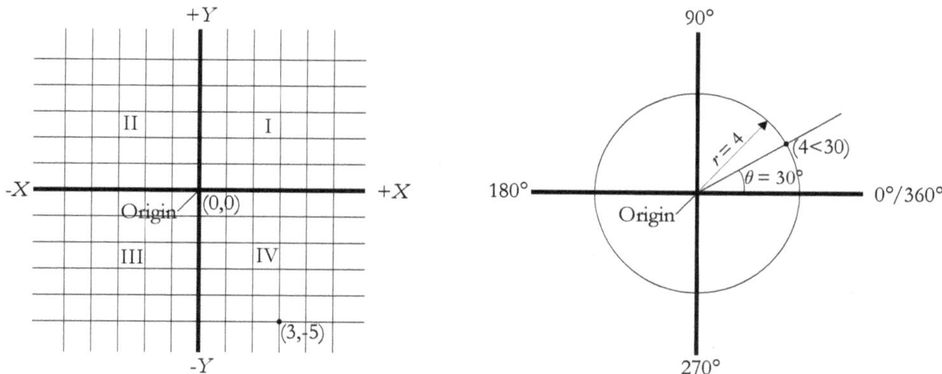

Figure 1.2 Cartesian coordinate system **Figure 1.3** Polar coordinate system

Polar Coordinate System

The polar coordinate system is a two-dimensional coordinate system in which each point on a plane is determined by a distance from a fixed point and an angle from a fixed direction. The fixed point (analogous to the origin of a Cartesian system) is called the pole, and the ray from the pole in the fixed direction is the polar axis. The distance from the pole is called the radial coordinate or radius, and the angle is the angular coordinate, polar angle. The input form of this system is $r<\theta$. Figure 1.3 shows the system, were 0° is a horizontal direction from left to right, 90° is straight up, 180° is horizontal from right to left, and so on. In this form, *r* represents the distance of the point from origin in the direction of θ angle. Angles are always measured in counterclockwise (anti-clockwise) direction and in the unit of degrees. You can also measure it in clockwise direction by giving a '–' (minus) sign, like -30. Hence, -30° angle represents 330°. A point at 4<30 means the point is situated at 4 units from the origin along the direction of 30° angle.

Concept of Relative Coordinate

The concepts of coordinate system, explained in the previous sections, are in absolute sense. To specify a coordinate in AutoCAD, one can use relative coordinates as well; either relative Cartesian or relative polar. The input forms of these systems are @*X*,*Y* and @$r<\theta$ respectively. The *at* symbol (@) tells AutoCAD that the displacement will be measured with reference to the previous point we have picked rather than to the origin. For example, at first we specify a point 2,2; now we are specifying another point by entering @1,2. In this case, the second point will be at a distance of 1 unit along the *X*-axis and 2 units along the *Y*-axis from 2,2 point (not from the origin). If we calculate from the origin, the point's absolute coordinate will be 3,4.

Drawing Lines

Draw toolbar: Line
Pull-down menu: Draw → Line
Command: LINE or L ↵

AutoCAD responds with the following prompt:

```
Specify first point:
```

It is asking for selecting a point in the drawing area to begin a line. If we move our mouse around the drawing area, we shall notice how the cross-hair cursor follows mouse movements. Try to select a point on the screen near the centre by clicking the pick button (left button) of the mouse. As you select the point, AutoCAD adds this in the command prompt area:

```
Specify next point or [Undo]:
```

Now move your mouse slowly and you will notice a line with one end fixed on your given point and the other end following the cursor. This is called *rubber-band line*. Now move the cursor near to top-right corner of the drawing area and press the pick button again. The first rubber-band line is now fixed between the two points you selected. A second rubber band line appears immediately. The AutoCAD adds the following in the prompt area again:

```
Specify next point or [Undo]:
```

Move your cursor and click again anywhere by the pick button of the mouse. The second line will appear on the screen between the second and third points you have selected. Now, AutoCAD adds the following in the prompt area:

```
Specify next point or [Close/Undo]:
```

AutoCAD draws a line segment and continues to prompt for points. One can draw a continuing series of line segments, but each line segment is a separate object. Press ENTER from keyboard to stop the command. Note that the line command terminates and returns the command prompt.

Notes:
1. In response to AutoCAD's prompt if we press enter ignoring AutoCAD is asking for, is called **null enter** or **null return**. It is a return without giving any input; may also be called as a *blank enter*.

2. Notice that the lines you have drawn are not smooth as you need. Lines are like stair. This is called **staircase effect**. It is just a resolution problem of the monitor. The lines will be perfect once printed.
3. One may use SPACEBAR instead of ENTER key. In AutoCAD, the function of *spacebar* is same as the *enter* key (except when you type a text in the drawing).
4. If you press enter in command prompt, AutoCAD repeats the previous command.

Now, let us try to draw a square of 1 unit side. Go through the line command again.

`Command:` **LINE ↵**
`Specify first point:` **1,1**
`Specify next point or [Undo]:` **@1<0** *(or 2,1; or @1,0)*
`Specify next point or [Undo]:` **@1<90** *(or 2,2; or @0,1)*
`Specify next point or [Close/Undo]:` **@1<180** *(or 1,2; or @-1,0)*
`Specify next point or [Close/Undo]:` **C ↵** *(C for close option)*

Congratulation! You have drawn the square as shown in Figure 1.4.

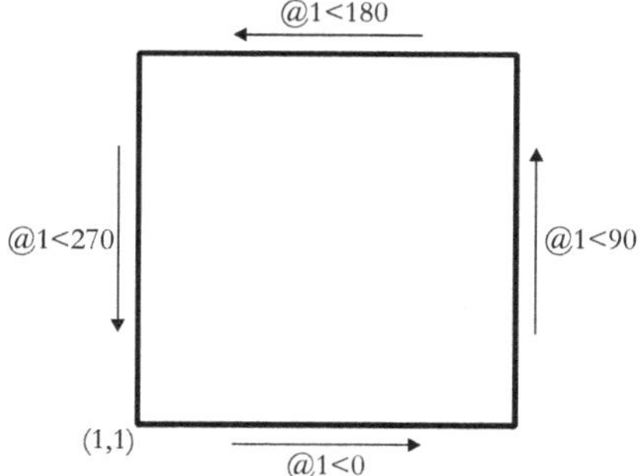

Figure 1.4 Distance and direction input for a square

Note: The close option can be used to join the current point with the initial point of the first line.

Erasing Objects

Modify toolbar: Erase
Pull-down menu: Modify → Erase
Command: ERASE or E ↵

After drawing some objects you may want to erase some of them from the screen. For this purpose, go through the ERASE command. When you invoke the erase command, AutoCAD responds with the prompt '`Select objects:`' and a small box (called *pick box*) replaces the screen cursor. To erase an object, move the pick box so that it touches the object and click the left mouse button. The object will be selected and '`Select objects:`' prompt returns again. You can either continue selecting objects or press null-enter to terminate the object selection process. The selected objects will be erased.

`Command:` **ERASE ↵**
`Select objects:` *Select 1st object*
`Select objects:` *Select 2nd object*
`Select object:` ↵

OOPS Command

Command: OOPS ↵

If you unintentionally or accidentally erased some objects, you can correct it by restoring the erased objects by means of OOPS command. Hence, OOPS can work just after the ERASE command and can restore only the objects erased by the latest ERASE command.

`Command:` **OOPS ↵**

Getting Out of Trouble

Beginners often make some mistakes. Before proceeding further, here are some powerful yet easy-to-use tools to help you to recover from accidental mistakes.

Escape [Esc] If you are in a command and you want to cancel or get out of that command, press Esc (Escape) key on the keyboard; the command terminates.

Undo If you accidentally have done some changes in the drawing and want to reverse that change, you can enter U in the command prompt. U stands for undo the last operation you have done. Each time you enter U, AutoCAD undoes one operation at a time in reverse order.

Drawing Circles

Draw toolbar: Circle
Pull-down menu: Draw → Circle
Command: CIRCLE or C ↵

To draw a circle you can use the CIRCLE command. The default option of CIRCLE command is *center-radius* option. If you want to draw a circle of 1 unit radius:

Command: **CIRCLE ↵**
Specify center point for circle or
[3P/2P/Ttr (tan tan radius)]: *Pick a point as center of the circle*
Specify radius of circle or [Diameter] <current>:**1 ↵**

> **Note:** In the above command <current> means the value of radius you have used previously, will be shown as current and default value of radius.

You can draw a circle by *center-diameter* option:

Command: **CIRCLE ↵**
Specify center point for circle or
[3P/2P/Ttr (tan tan radius)]: *Pick a point as center of the circle*
Specify radius of circle or [Diameter] <current>:**D ↵** *(D for entering the diameter option)*
Specify diameter of circle <current>: **2 ↵**

> **Note:** If you want to use an option other than the default option in a command, you have to type the option or only the upper case letters of the option, and press ENTER. Some examples are D for Diameter, T for Ttr, LA for LAyer, etc.

You may also want to draw a circle by *two-point* option. These two points are only the diametrical end points as shown in Figure 1.5.

Command: **CIRCLE ↵**
Specify center point for circle or
[3P/2P/Ttr (tan tan radius)]: **2P ↵**
Specify first end point of circle's diameter: *Pick a point*
Specify second end point of circle's diameter: *Pick another point*

If you want to draw a circle using *three-point* option, following steps to be followed:

Command: **CIRCLE ↵**
Specify center point for circle or
[3P/2P/Ttr (tan tan radius)]: **3P ↵**

`Specify first point on circle:` *Pick a point*
`Specify second point on circle:` *Pick a point*
`Specify third point on circle:` *Pick a point*

Remember, the given points should not be collinear.

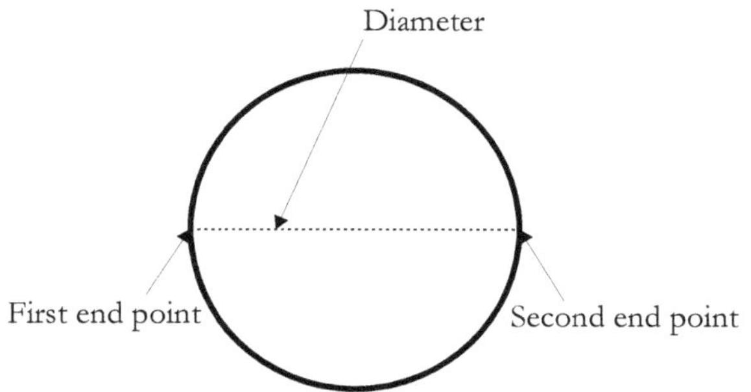

Figure 1.5 Drawing a circle using 2P option

First tangent
Second tangent
Radius = 1

Figure 1.6 TTR option of circle

You can use the Tangent-Tangent-Radius (TTR) option (Figure 1.6). A tangent is an object (line, circle or arc) that contacts the circumference of a circle at only one point. In the TTR option you have to select two objects which are to be tangents to the circle. Then you have to specify the radius of the circle. The prompt sequence is as follows:

`Command:` **CIRCLE ↵**
`Specify center point for circle or`
`[3P/2P/Ttr (tan tan radius)]:` **T ↵**
`Specify point on object for first tangent of circle:` *Select an object which will be tangent to the circle*
`Specify point on object for second tangent of circle:` *Select another object which will be the second tangent to the circle*
`Specify radius of circle <current>:` *Enter desired radius of the circle*

Drawing Doughnuts

Pull-down menu: Draw → Donut
Command: DO or DONUT or DOUGHNUT ↵

The DONUT or DOUGHNUT command is issued to draw an object that looks like a filled circular ring as shown in Figure 1.7.

`Command:` **DO ↵**

```
Specify inside diameter of donut <current>: Specify the inner diameter
Specify outside diameter of donut <current>: Specify the outer dia
Specify center of donut or <exit>: Specify the center point of doughnut
Specify center of donut or <exit>: Specify the center point of doughnut to
    draw more doughnuts of previous specification or give a null enter to exit the command
```

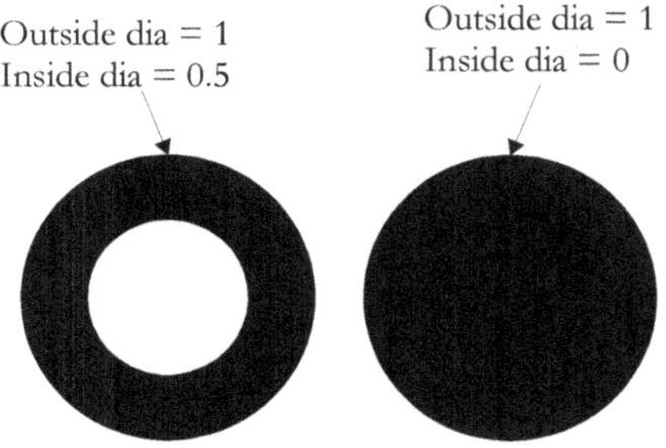

Figure 1.7 Doughnuts

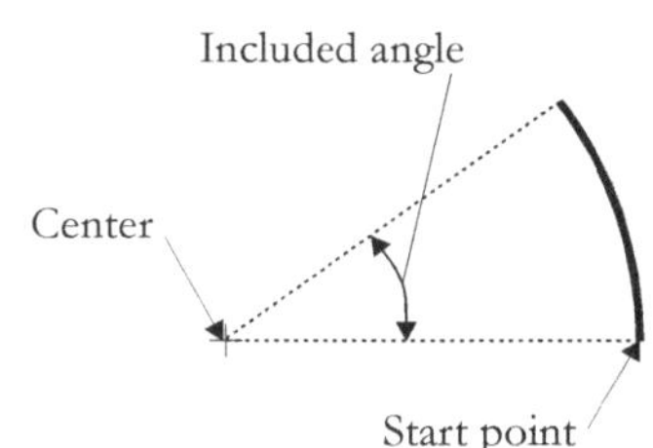

Figure 1.8 The *start-center-angle* option

Drawing Arcs

Draw toolbar: Arc
Pull-down menu: Draw → Arc
Command: ARC or A ↵

An arc is defined as a part of a circle. The default option of ARC command is *three-point* option. The three-point option requires the start point, second point and the end point of the arc. Given points should not be collinear.

```
Command: ARC ↵
Specify start point of arc or [Center]: Pick a point
Specify second point of arc or [Center/End]: Pick a point
Specify end point of arc: Pick a point
```

Now, let us try to use the *start-center-angle* option. This option is useful if you know the included angle of arc as shown in Figure 1.8.

```
Command: ARC ↵
Specify start point of arc or [Center]: Pick a point
Specify second point of arc or [Center/End]: C ↵
Center: Pick a point as center point of arc
Specify end point of arc or [Angle/chord Length]: A ↵
Specify included angle: Enter the value of included angle
```

Remember, the arc will be drawn in counterclockwise direction. You can also draw in clockwise direction by entering a '-' (negative sign) followed by the value of included angle.

AutoCAD commands work within a distinct structure. You first issue a command, which in turns offers options in the form of a prompt. Depending on the options you have selected, you will get another set of options or you will be prompted to take some action, such as picking a point, selecting an object, or entering a value. You will become intimately familiar with this routine as you continue to work through exercises. Once you understand the workings of command prompts and dialog boxes, you can almost teach yourself the rest of the program.

An illustration of ARC command and its options are shown in Figure 1.9. This will help you to explore the other options of ARC command.

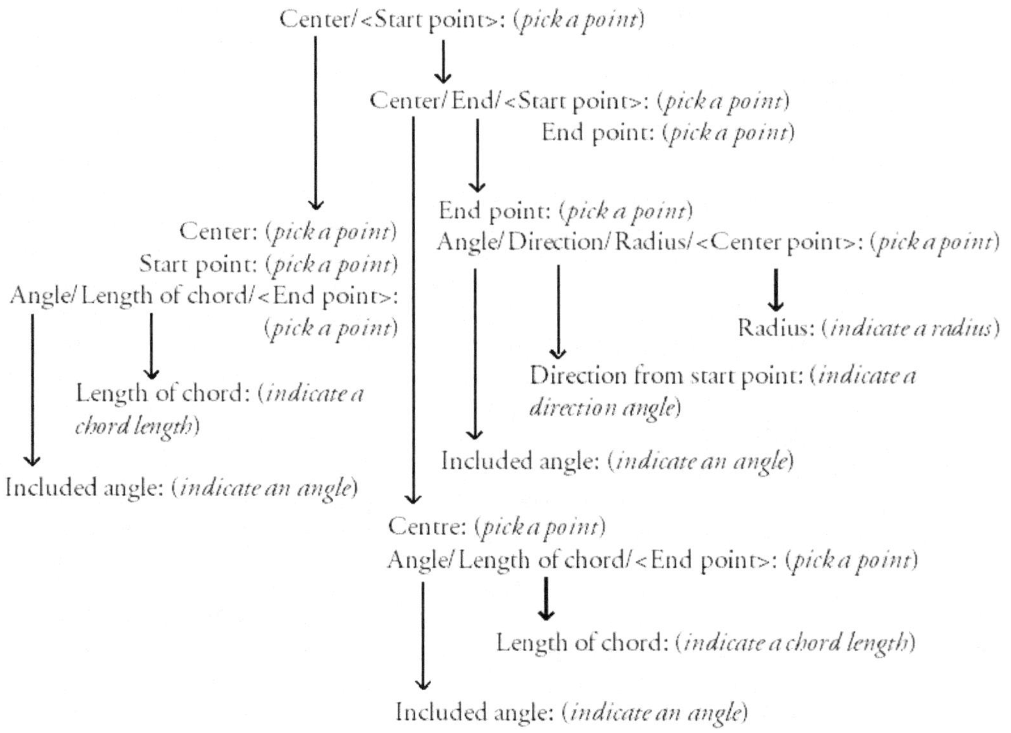

Figure 1.9 Typical structure of ARC command

Drawing Rectangles

Draw toolbar: Rectangle
Pull-down menu: Draw → Rectangle
Command: RECTANG or REC ↵

A rectangle can be drawn using RECTANGLE command (Figure 1.10). After invoking the RECTANGLE command, you are prompted to specify the diagonal corners of the rectangle.

```
Command: REC ↵
Specify first corner point or [Chamfer/Elevation/Fillet/
Thickness/Width]: specify a point
Specify other corner point or [Area/Dimensions/Rotation]:
                                   Specify diagonally opposite corner
```

You also have the options to enter *area*, or *dimensions*, or *rotation*. Area option creates a rectangle using the area and either a length or a width. If the Chamfer or Fillet option is active, the area includes the effect of the chamfers or fillets on the corners of the rectangle. Dimensions option creates a rectangle using length and width values. Rotation option creates a rectangle at a specified rotation angle.

The RECTANGLE command offers the following options (Figure 1.10):

Chamfer: The chamfer option creates a chamfer by specifying the chamfer distances.

Fillet: The fillet option creates a filleted rectangle by specifying the radius.

Width: The width option allows you to control the line width of the rectangle by specifying the width.

Thickness and *Elevation* options are applicable for 3D drawing.

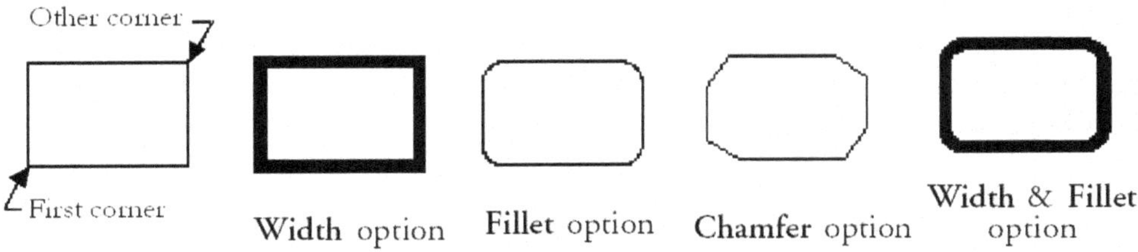

Figure 1.10 Using RECTANGLE command

Drawing Regular Polygons

Draw toolbar: Polygon
Pull-down menu: Draw → Polygon
Command: POLYGON or POL ↵

Polygon is a closed geometric figure with equal sides and equal angles. The number of sides may vary from 3 to 1024. The default option is as follows:

`Command:` **POLYGON ↵**
`Enter number of sides <4>:` **6 ↵** *(4 for square, 5 for pentagon, etc.)*
`Specify center of polygon or [Edge]:` *Specify the center of polygon*
`Enter an option [Inscribed in circle/Circumscribed about circle] <I>:` ↵ *(Press ENTER for Inscribed option or C for Circumscribed)*
`Specify radius of circle:` *Enter the radius of circle or drag the mouse and click*

Inscribed and Circumscribed polygons are shown in Figure 1.11. If you select a point to specify the radius of an inscribed polygon, one of the vertices is positioned on the selected point. In the case of circumscribed polygons, the midpoint of an edge is placed on the point you have specified. In the case of numerical specification of the radius, the bottom edge of the polygon will be paralleled to *X*-axis.

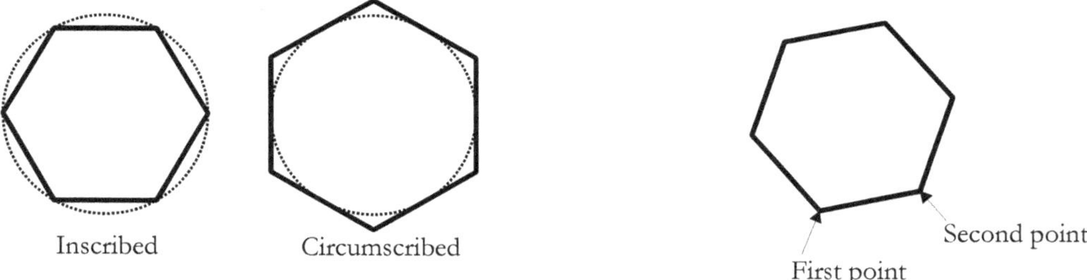

Figure 1.11 Inscribed and circumscribed polygons **Figure 1.12** Polygon using *edge* option

The other method for drawing a polygon is to select the *edge* option. For this option, you need to specify the two endpoints of an edge of the polygon. The polygon is drawn in counterclockwise direction, with the two points entered defining its first edge and the angle of first edge with *X*-axis (Figure 1.12).

`Command:` **POLYGON ↵**
`Enter number of sides <4>:` **6 ↵**
`Specify center of polygon or [Edge]:` **E ↵**
`Specify first endpoint of edge:` *Specify first point*
`Specify second endpoint of edge:` *Specify second point*

Drawing Polylines

Draw toolbar: Polyline
Pull-down menu: Draw → Polyline
Command: PLINE or PL ↵

A polyline is a line that can have different characteristics. One of the features of polyline is: polylines are thick lines having a desired width. A single polyline object can be formed by joining polylines and polyarcs of different thickness. A polyline may have multiple segments; however, all segments act like a single object. The PLINE command functions fundamentally like the LINE command.

Command: **PLINE ↵**
Specify start point: Specify the starting point of polyline
Current line width is 0.0000
Specify next point or [Arc/Halfwidth/Length/Undo/Width]:
Specify the endpoint of first polyline segment
Specify next point or [Arc/Close/Halfwidth/Length/Undo/Width]:
Specify the endpoint of the second polyline segment, or press null enter to exit the command

> **Note:** All segments, drawn by a single PLINE command, are a single entity. The rectangles drawn by RECTANG command and polygons drawn by POLYGON command are basically closed polylines.

You can change the current polyline width by entering *width* option. Then you will be prompted for the starting width and the ending width of the polyline. Here is the command sequence for an example:

Command: **PLINE ↵**
Specify start point: **2,2 ↵**
Current line width is 0.0000
Specify next point or [Arc/Halfwidth/Length/Undo/Width]: **W ↵**
Specify starting width <0.0000>: **0.1 ↵**
Specify ending width <0.1000>: **0.05 ↵**
Specify next point or [Arc/Halfwidth/Length/Undo/Width]: **@2,0 ↵**
Specify next point or [Arc/Close/Halfwidth/Length/Undo/Width]: **W ↵**
Specify starting width <0.1000>: **0 ↵**
Specify ending width <0.0000>: **0.2 ↵**
Specify next point or [Arc/Close/Halfwidth/
Length/Undo/Width]: **@1<90 ↵**
Specify next point or [Arc/Close/Halfwidth/Length/Undo/Width]: ↵

Similarly, you can set the halfwidth (half width) of a polyline segment with the help of *halfwidth* option. The *length* option prompts you to enter the length of a new polyline segment. The new polyine segment will be in the length you have entered. It will be drawn at the same angle as the last segment. The *undo* option erases the most recently drawn polyline segment. The *close* option closes the polyline by drawing a polyline segment from the most recent endpoint to the initial start point. The close and undo options of polyline are just like the close and undo options of LINE command respectively.

Another option of polyline is *arc* option. This option is used to switch from drawing polylines to polyarcs, and provides you the options associated with drawing polyarcs. The arc option can be invoked by entering A at the following prompt:

```
Specify next point or [Arc/Close/Halfwidth/Length/Undo/Width]: A ↵
Specify endpoint of arc or [Angle/CEnter/CLose/Direction/
Halfwidth/Line/Radius/Second pt/Undo/Width]: Draw arc using the options
```

The options you have used in ARC command are available here and some options of PLINE also. If you want to come out into the line mode from arc mode you have to type L at the following prompt:

```
Specify endpoint of arc or[Angle/CEnter/CLose/Direction/
Halfwidth/Line/Radius/Second pt/Undo/Width]: L ↵
Specify next point or [Arc/Close/Halfwidth/Length/Undo/Width]:
```

Drawing Points

Draw toolbar: Point

Pull-down menu: Draw → Point → Single Point or Multiple Point

Command: POINT or PO ↵

To draw a point anywhere on the screen, AutoCAD provides the POINT command. If you invoke the POINT command by entering POINT at the command prompt, you can draw only one point by a single POINT command. On the other hand, if you invoke the POINT command from the toolbar or the pull-down menu (Multiple Point), you can draw as many points as you desire by a single command. In this case you can exit from the POINT command by pressing Esc key.

```
Command: POINT ↵
Specify a point: Specify the location of point
```

Changing the Point Style

Pull-down menu: Format → Point Style....
Command: DDPTYPE ↵

When you invoke the command, the Point Style dialog box will appear on the screen (Figure 1.13). There are 20 combinations of point types. You can select a point style in the dialog box by clicking on the point style of your choice. Now all the points will be drawn with the selected style until you change it to a new style. You can also set point style by entering PDMODE at the Command prompt and changing its value to that of the required point type.

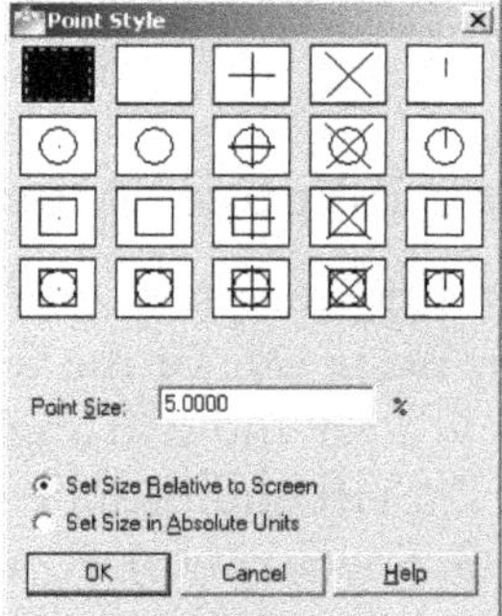

Figure 1.13 Point Style dialog box

Drawing Ellipse

Draw toolbar: Ellipse
Pull-down menu: Draw → Ellipse
Command: ELLIPSE or EL ↵

If a circle is observed from an angle, the shape seen is called an ellipse. Once you invoke the ELLIPSE command, AutoCAD will acknowledge with the following prompt:

```
Command: ELLIPSE ↵
Specify axis endpoint of ellipse or [Arc/Center]: 3,3 ↵
Specify other endpoint of axis: 4,2 ↵
Specify distance to other axis or [Rotation]: 1 ↵
```

If you enter R (rotation) at the '`Specify distance to other axis or [Rotation]:`' prompt, the first axis specified is automatically taken as the major axis of the ellipse. The next prompt is '`Rotation around major axis`'. The major axis is taken as the diameter line of the circle, and the rotation takes around the diameter line into the third dimension.

```
Command: ELLIPSE ↵
Specify axis endpoint of ellipse or [Arc/Center]: Specify a point
Specify other endpoint of axis: Specify the other end point
Specify distance to other axis or [Rotation]: R ↵
```

`Specify rotation around major axis:` *Specify your desired rotation angle*

If you want to draw an ellipse by using the *center and two axis* option, you can construct it by specifying the center point, the endpoints of one axis, and the length of other axis.

Drawing Elliptical Arcs

You can draw elliptical arcs by using ELLIPSE command. You can also define the arc limits by using one of the following options:

1. Start and end angles of the arc
2. Start and included angles of the arc
3. Specifying start and end parameters

Let us consider the following examples:

a. Start angle -45, end angle 135 (Figure 1.14)
b. Start angle -45, included angle 225 (Figure 1.15)
c. Start parameter @1,0 and end parameter @1<225 (Figure 1.16)

a. `Command:` **ELLIPSE ↵**
`Specify axis endpoint of ellipse or [Arc/Center]:` **A ↵**
`Specify axis endpoint of elliptical arc or [Center]:` *Pick a point*
`Specify other endpoint of axis:` *Pick the other end point*
`Specify distance to other axis or [Rotation]:` *Pick another point*
`Specify start angle or [Parameter]:` **-45 ↵**
`Specify end angle or [Parameter/Included angle]:` **135 ↵** (*Angle where the arc ends*)

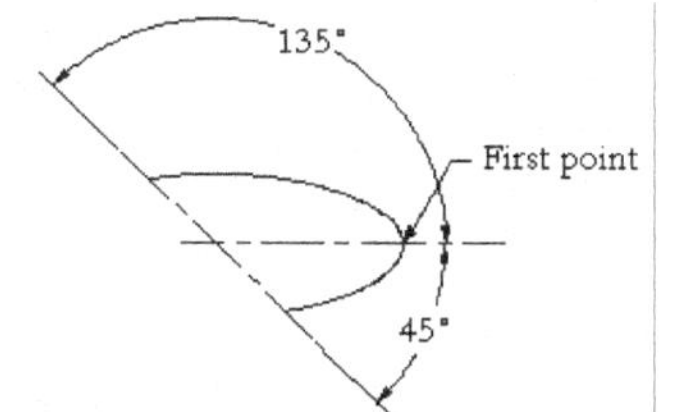

Figure 1.14 Elliptical arc using start angle and end angle

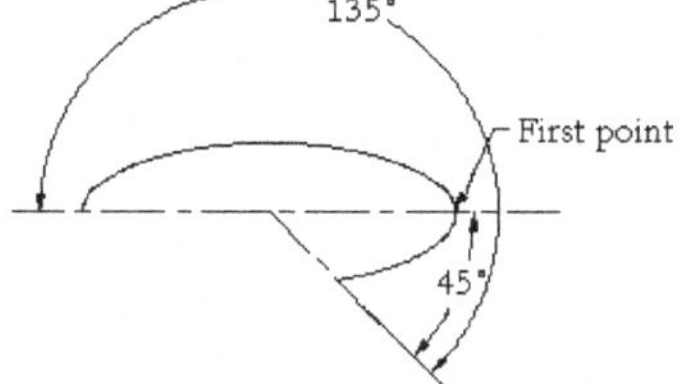

Figure 1.15 Elliptical arc using start angle and included angle

b. `Command:` **ELLIPSE ↵**
`Specify axis endpoint of ellipse or [Arc/Center]:` **A ↵**
`Specify axis endpoint of elliptical arc or [Center]:` *Pick a point*
`Specify other endpoint of axis:` *Pick the other end point*
`Specify distance to other axis or [Rotation]:` *Specify another point*

```
Specify start angle or [Parameter]: -45 ↵
Specify end angle or [Parameter/Included angle]: I ↵
Specify included angle for arc <180>: 225 ↵
```

c.

```
Command: ELLIPSE ↵
Specify axis endpoint of ellipse or [Arc/Center]: A ↵
Specify axis endpoint of elliptical arc or [Center]: Pick a point
Specify other endpoint of axis: Pick the other end point
Specify distance to other axis or [Rotation]: Specify another point
Specify start angle or [Parameter]: P ↵
Specify start parameter or [Angle]: @1,0 ↵
Specify end parameter or [Angle/Included angle]: @1<225 ↵
```

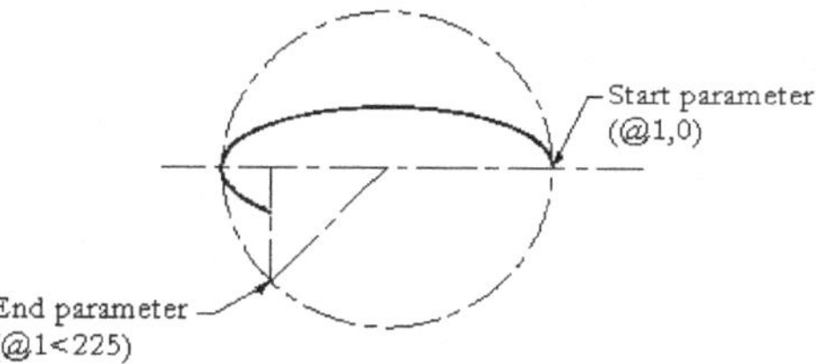

Figure 1.16 Elliptical arc using Start & End parameters

Writing Text

Pull-down menu: Draw → Text → Single Line Text
Command: DTEXT or DT ↵

DTEXT stands for dynamic text. This command is very useful to write some textual part on your drawing.

```
Command: DTEXT ↵
Current text style: "Standard" Text height: 0.200 Annotative: No
Specify start point of text or [Justify/Style]: Specify the start point
Specify height <0.2000>: Enter desired height of the text to be written
Specify rotation angle of text<0>: Enter the angle at which the text will be written
Enter text: Type the first line of your text and press ENTER
Enter text: Type the second line or NULL ENTER for none
```

AutoCAD offers different options for aligning the text. Alignment refers to the layout of the text. The prompt sequence for using these options:

```
Command: DTEXT ↵
```

Specify start point of text or [Justify/Style]: **J ↵**
Enter an option [Align/Fit/Center/Middle/Right/TL/TC/TR/
ML/MC/MR/BL/BC/BR]: **C ↵**
Specify center point of text: *Specify the center point of the text*
Specify height <current>: *Enter the desired height of the text*
Specify rotation angle of text <current>: *Enter the angle*
Enter text: *Type the first line of your text and press ENTER*
Enter text: *Type the second line or NULL ENTER for none*

Other justify options are shown in Figure 1.17. With the *style* option you can change the existing text style. We shall discuss more about the text style later in this chapter.

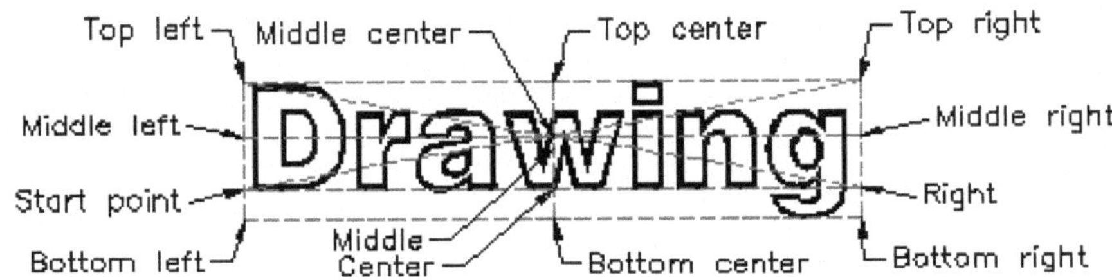

Figure 1.17 Justify options of DTEXT command

Special characters

Almost in all drafting applications, you need to draw special characters (symbols) in the normal text and in the dimension text; for example, degree symbol or the diameter symbol. These can be written with the appropriate sequence of control characters (control code). The control codes for some of the symbols are as follows:

Control code	*Special characters*
%%c	Diameter symbol
%%d	Degree symbol
%%p	Plus/minus symbol
%%o	Toggle for overscore mode on/off
%%u	Toggle for underscore mode on/off

Creating Text Style

Pull-down menu: Format → Text Style...
Command: STYLE or ST ↵

You can change the text style or can create a new style using STYLE command. When you enter the command, the Text Style dialog box (Figure 1.18) will be displayed.

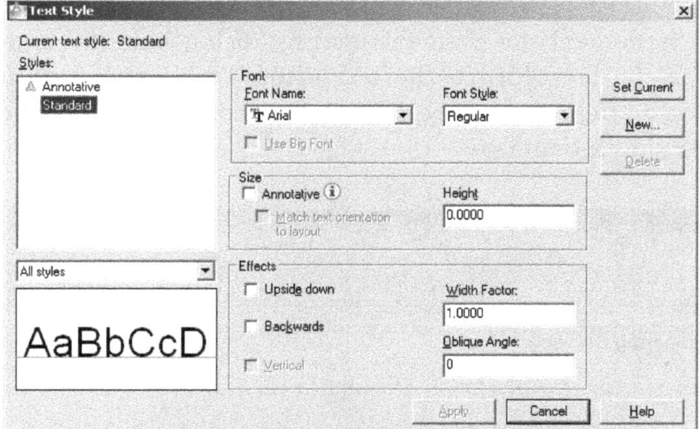

Figure 1.18 Text Style dialog box

Following are the steps for setting a new text style:

1. Click on **New...** for a new style. New Text Style dialog box will appear.
2. Enter the name of the style you want to create and click on **OK** in the New Text Style dialog.
3. Select the **Font Name** against new style. You can see the preview of the style on the lower right corner of Text Style dialog box.
4. You can change the **Width Factor**, **Oblique Angle**, **Height** and other **Effects** also. If you set the text height as 0, then AutoCAD will ask for the height in the DTEXT command and you can change it every time. But if once you set the height you will not be prompted for text height in DTEXT command.
5. Click on **Apply**.
6. Now, create another text style or **Close** the dialog box.

Multiline Text

Draw toolbar: A Multiline Text

Pull-down menu: Draw → Text → Multiline Text

Command: MTEXT or MT ↵

You can use the MTEXT command to write a paragraph whose width can be specified by defining two corners of the text boundary or by entering a width.

```
Command: MTEXT ↵
Current text style: "Standard" Text height: 0.2000 Annotative: No
```

```
Specify first corner: Select a point to specify the first corner
Specify opposite corner or [Height/Justify/Line spacing/
Rotation/Style/Width/Columns]: Select other corner for specifying the width of text
```

Once the width is defined the Text Formatting dialog box will appear as shown in Figure 1.19. Type the text, format the text using Text Formatting tools, and finally click on OK. The text created by the MTEXT command is a single object regardless of the number of lines it contains.

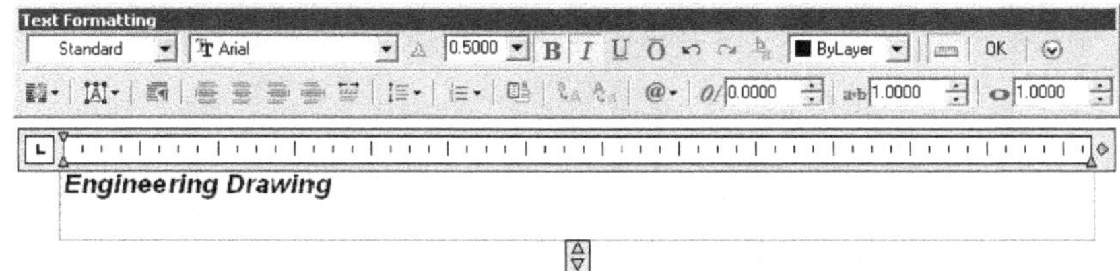

Figure 1.19 Text Formatting dialog box

Hatching

Draw toolbar: Hatch
Pull-down menu: Draw → Hatch...
Command: BHATCH or H ↵

Filling the objects with a pattern is known as hatching. This hatching process can be accomplished by using the BHATCH command. The BHATCH (boundary hatch) command allows you to hatch a region enclosed within a boundary (closed area). After invoking BHATCH command the Hatch and Gradient dialog box (Figure 1.20) will appear on the screen. Follow the sequence below:

1. Click on the **Pattern...** button. The Hatch Pattern Palette dialog box will appear (Figure 1.21).
2. Select your desired hatch pattern and click on **OK**. The Hatch Pattern Palette dialog will disappear and you will back to Hatch and Gradient dialog box.
3. Click on **Add: Pick points** button and select a point inside the boundary to be hatched. The other option is clicking on **Add: Select objects** button to select the objects to be hatched.
4. Press NULL ENTER at the command prompt. The Hatch Pattern Palette dialog will reappear.
5. Click on **Preview** button. This allows you to preview the hatch before actually applying it.
6. Now, press the Esc key to escape from the preview mode.

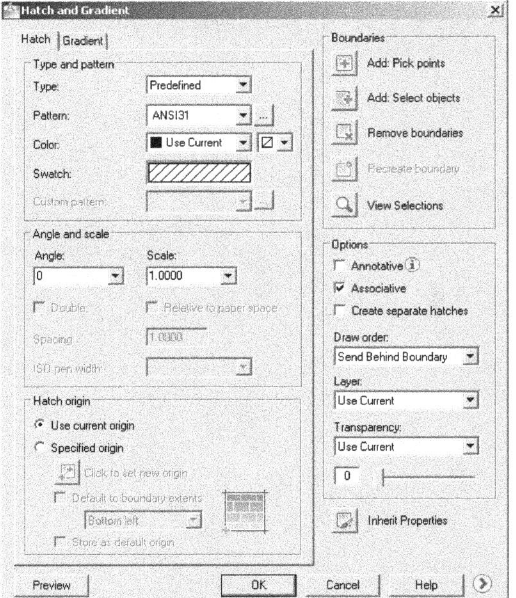

Figure 1.20 Hatch and Gradient dialog box

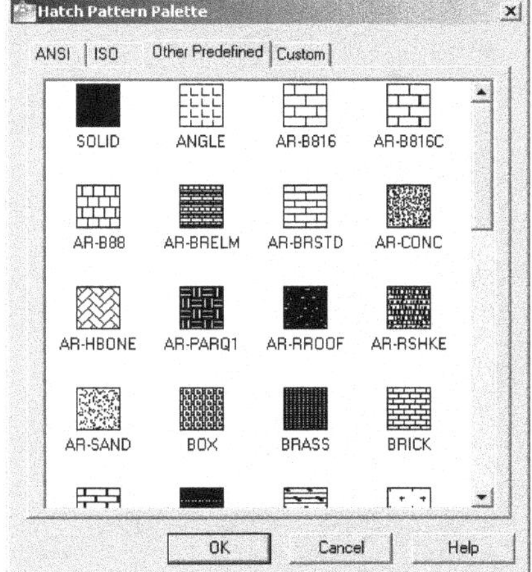

Figure 1.21 Hatch Pattern Palette dialog box

7. If you are not satisfied with the density of hatch, you can control the density by changing the **Scale**. Further, you can specify an angle by entering a value in the **Angle** option.
8. Go for preview again; and if you are satisfied go for applying the hatch by pressing the ENTER key or by clicking the OK button in the Hatch and Gradient dialog.

Hatching lines are associative. Hatch drawn by a command is a single entity. You can also draw exploded hatch turning off the **Associative** option in the Hatch and Gradient dialog box. An associative hatch can be exploded by the command **EXPLODE** at the command prompt.

Boundary Style

Click on More Options button in the Hatch and Gradiant dialog box (Figure 1.20); several other options will appear. Among them, Island detection (Figure 1.22) is very important. There are three styles from which you can choose one: Normal, Outer, or Ignore.

Normal: This style hatches inward starting at the outermost boundary. If it encounters an internal boundary, it turns off the hatching. In this manner, alternate areas are hatched, starting with the outermost area.

Outer: This option hatches only the outer most boundary. It turns off hatching all the inner boundaries.

Ignore: In this option, all areas bounded by the outermost boundary are hatched, ignoring any hatch boundaries that are within the outer boundary.

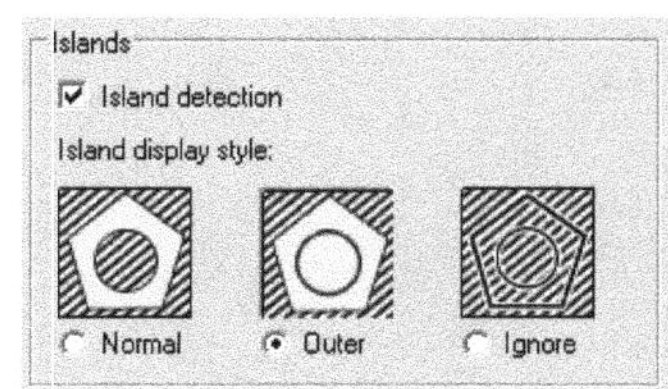

Figure 1.22 Island detection options

Object Snap

Toolbar: Object Snap

Object Snap is one of the most useful features of AutoCAD. The term OBJECT SNAP refers to the cursor's ability to snap exactly to a geometric point on an object. For example, if you want to place a point at the midpoint of a line, you may not be able to specify the exact point. Using the *midpoint* object snap, it will be possible. You can use the Object Snap toolbar (Figure 1.23) or cursor menu (Figure 1.24) for this purpose. You can access the cursor menu by pressing the middle button of a three-button mouse or holding down the Shift key on the keyboard and then pressing the right button of your two-button mouse.

Some frequently used object snap modes are (uppercase letters can be used at the command prompt for a specific snap mode):

NEArest The nearest object snap mode selects a point on an object (line, arc, circle etc.) that is visually closest to the graphic cursor.

ENDpoint The endpoint object snap mode selects closest endpoint of a line or an arc.

MIDpoint The midpoint object snap mode selects mid point of a line or an arc.

INTersection The intersection object snap mode selects a point where two or more objects intersect.

CENter The center object snap mode selects the center point of an ellipse, circle or arc.

TANgent The tangent object snap allows you to draw a tangent to or from an existing ellipse, arc or circle.

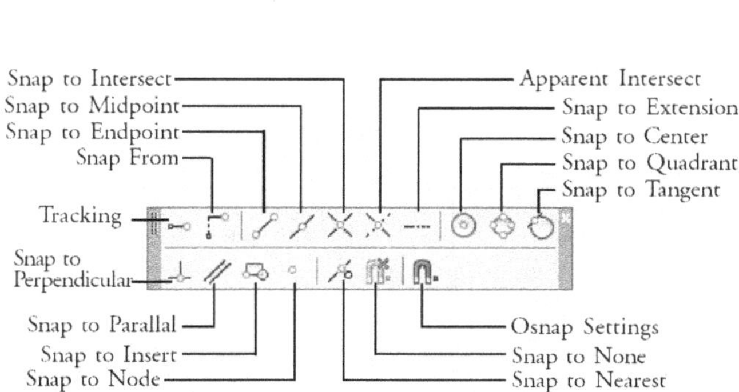

Figure 1.23 Object Snap toolbar

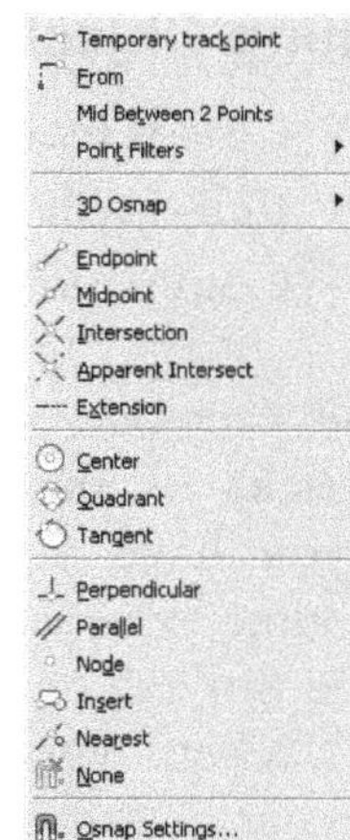

Figure 1.24 Object Snap cursor menu

QUAdrant The quadrant object snap mode selects a quadrant point of an ellipse, circle or arc.

PERpendicular The perpendicular object snap mode is used to draw a line perpendicular to or from another line / normal of an arc or circle.

NODe The node object snap mode selects a point drawn by a POINT command.

INSert The insert object snap mode selects an insertion point of text, block etc.

NONE The none object snap mode turns off any running object snap.

TRAcking The tracking can be used to locate points with respect to another point. It helps users to select orthogonal points relative to another point.

To practice with the snap modes, first draw a line. Then, try to draw a circle centering one of the endpoints of the previously drawn line. Follow the stated command sequence:

`Command:` **CIRCLE ↵**
`Specify center point for circle or`
`[3P/2P/Ttr (tan tan radius)]:` **END ↵**
`of` *Move the cursor at one of the end points of the previously drawn line, the end point of the line will be highlighted, now click the left mouse button. The end point of the line will be considered as the center of the circle.*
`Specify radius of circle or [Diameter]:` *Enter desired value*

Similarly other snap modes can also be used. Try to use them extensively.

Running Object Snap

Cursor menu: Osnap Settings...
Pull-down menu: Tools → Drafting Settings...
Command: OSNAP or OS ↵

In the previous section you have learned how to use object snaps. One of the drawbacks of the described method is that you have to select a snap mode every time you use them. This problem can be solved by using running object snap. When you enter the OSNAP command, Drafting Settings dialog box appears on the screen as shown in Figure 1.25. You can set the running object snap modes in Object Snap tab by selecting the boxes (check boxes) before the snap modes. Once you set the running object snap mode, you are automatically in the mode and the marker is displayed when you move over the snap points. You can turn the running object snap ON or OFF by pressing F3 key from keyboard.

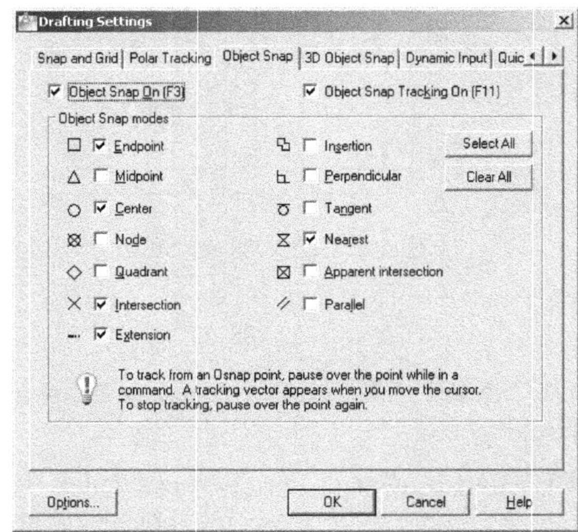

Figure 1.25 Drafting Settings dialog box

Saving Your Drawing

Standard toolbar: Save
Pull-down menu: File → Save
Command: QSAVE ↵

Saving your work is perhaps the most important task. In the AutoCAD, when you invoke the QSAVE command the Save Drawing As dialog box (Figure 1.26) will appear for the first time. Following are the steps to be followed in order to save a file:

1. Click on **Save in** combo box in the Save Drawing As dialog. Select your desired folder.
2. Go to **File name** edit box. Type the desired name of your drawing (file name Drawing1.dwg will be there by default, change the name according to your requirement).
3. Click on **Save** button.

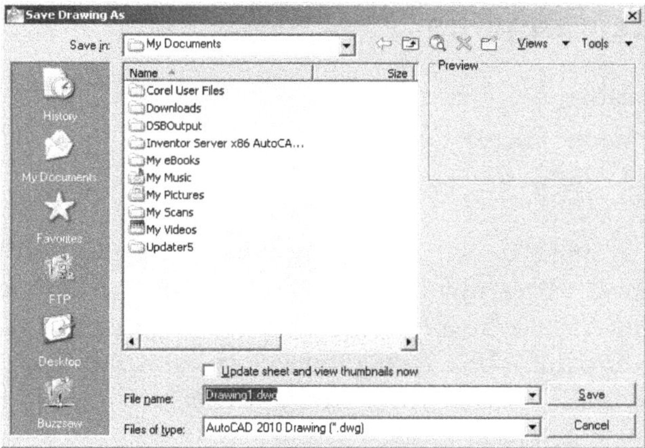

Figure 1.26 Save Drawing As dialog box

Creating a New File

Standard toolbar: New
Pull-down menu: File → New...
Command: NEW ↵

The NEW command is used to create a new drawing. After you invoke the new command, Select Template dialog box (Figure 1.27) will appear, in which the following are the steps to be followed:

1. Select **Acad.dwt** file from the template list.
2. Click on **Open** button.

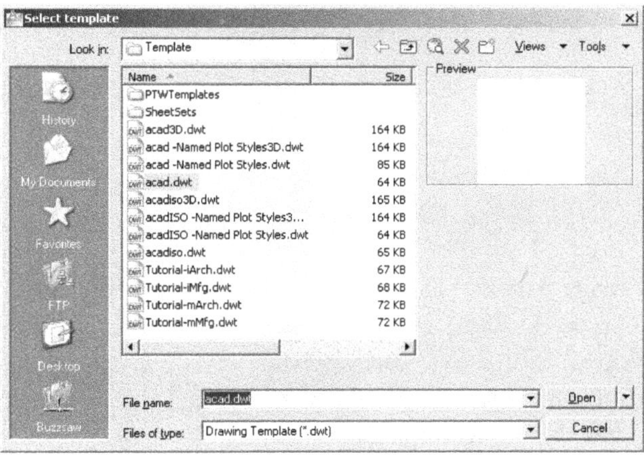

Figure 1.27 Select Template dialog box

Opening an Existing File

Standard toolbar: Open
Pull-down menu: File → Open...
Command: OPEN ↵

If you want to see or edit a drawing previously drawn by you, you have to go through the OPEN command. After invoking the OPEN command Select File dialog box (Figure 1.28) will appear; in which Following are the steps to be followed:

1. Click on **Look in** combo box. Select the folder where you have saved the drawing.
2. Now select the file which one you want to open. You can see preview of the drawing in Preview image box.
3. If the selection is correct, click on **Open** button.

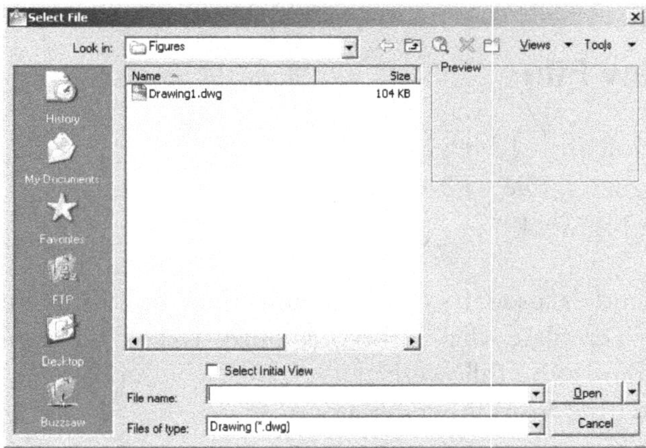

Figure 1.28 Select File dialog box

Quitting AutoCAD

Pull-down menu: File → Exit
Command: QUIT ↵ (or EXIT or END)

If you want to exit from AutoCAD just enter the QUIT command. If you have drawn or edited anything in a drawing and then used the command NEW or OPEN or QUIT without saving it, AutoCAD allows you to save the work first through a dialog box as shown in Figure 1.29. If you want to save the work, click on Yes button, otherwise click on No button.

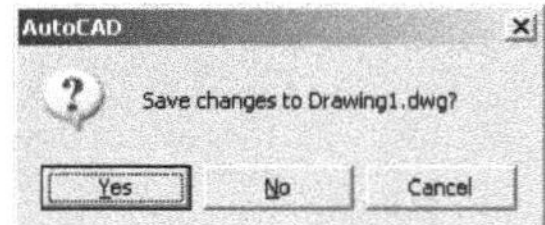

Figure 1.29 Confirmation dialog box

Some Important Tips

Using SPACEBAR as ENTER

To enter a command by using the keyboard, type the full command name or alias on the command line and press ENTER or SPACEBAR, or right-click your pointing device.

Command Aliases

Frequently used commands have abbreviated names. For example, instead of entering circle to start the CIRCLE command, you can enter C. Abbreviated command names are called *command aliases* and are listed in the acad.pgp file. To access the acad.pgp, on the **Tools** menu, click **Customize → Edit Program Parameters (acad.pgp)**.

Specifying Command Options

When you enter commands on the command line, AutoCAD displays either a set of options or a dialog box. For example, when you enter circle at the command prompt, the following prompt is displayed:

```
Specify center point for circle or [3P/2P/Ttr (tan, tan, radius)]:
```

You can specify the center point either by entering *X,Y* coordinate values or by using the pointing device to click a point on the screen.

To choose a different option, enter the letters capitalized in one of the options in the brackets. You can enter uppercase or lowercase letters. For example, to choose the three-point option (3P), enter 3p.

Executing a Command

To execute commands, press SPACEBAR or ENTER, or right-click your pointing device after entering command names or responses to prompts. Assume this step because AutoCAD do not specifically instruct you to press ENTER after each entry.

Transparent Commands

Many commands can be used transparently; i.e., they can be entered on the command line while you use another command. Transparent commands frequently change drawing settings or display options, for example, CAL, FILTER, PAN, OSNAP, GRID or ZOOM. In the command reference, transparent commands are designated by an apostrophe (') in front of the command name.

To use a command transparently, use the toolbar button or enter an apostrophe (') before entering the command at any prompt. On the command line, double angle brackets (>>) precede prompts that AutoCAD displays for transparent commands. After you complete the transparent command, the original command resumes. In the following example, you can zoom-in while you draw a line, and then you continue drawing the line.

```
Command: LINE ↵
Specify first point: 'ZOOM ↵
>>Specify corner of window, enter a scale factor (nX or nXP), or
[All/Center/Dynamic/Extents/Previous/Scale/Window] <real time>:
                                     Specify the first point of the window
>>>>Specify opposite corner: Specify the opposite corner of the window
Resuming LINE command
Specify first point:
```

Commands that do not select objects, or create new objects, or end the drawing session usually can be used transparently. Changes made in dialog boxes that you have opened transparently cannot take effect until the interrupted command has been executed. Similarly, if you reset a system variable transparently, the new value cannot take effect until you start the next command.

UNDO and REDO

With AutoCAD, you can backtrack your recent actions and also you can reverse the effect as you need.

Undo a Single Action: The simplest method of backtracking is to use Undo tool in the Standard toolbar or the U at the command prompt to undo a single action. Many commands include their own undo option so that you can correct mistakes without exiting the command. When you are creating lines and polylines, for example, enter U to undo the last segment.

Undo Several Actions at Once: Use the *mark* option of UNDO to mark an action as you work. You can then use the *back* option of UNDO to undo all actions that occurred after the marked action. Use the *begin* and *end* options of UNDO to undo actions you have defined as a group. You can also undo several actions at once with the Undo list in the Standard toolbar.

Reverse the Effect of Undo: You can reverse the effect of a single U or UNDO command by using REDO immediately after using U or UNDO. You can also redo several actions at once with the Redo list in the Standard toolbar.

2
Editing

Introduction

To use AutoCAD effectively, you need to know the editing commands and how to use them. In this section you will learn about the editing commands. We have discussed ERASE and OOPS command in Chapter 1. Now we are going to discuss other editing commands.

MOVE Command

Modify toolbar: Move
Pull-down menu: Modify → Move
Command: MOVE or M ↵

Sometimes objects may not be located where they should be. In such cases you can use MOVE command. This command lets you move the objects from their present location to a new one (Figure 2.1).

`Command:` **MOVE ↵**
`Select objects:` *Select the object you want to move*
`Select objects:` *Select another object or press ENTER*
`Specify base point or [Displacement] <Displacement>:` *Specify a point on or near to the object*

```
Specify second point of displacement or
<use first point as displacement>:
```
Select the new location by specifying a point

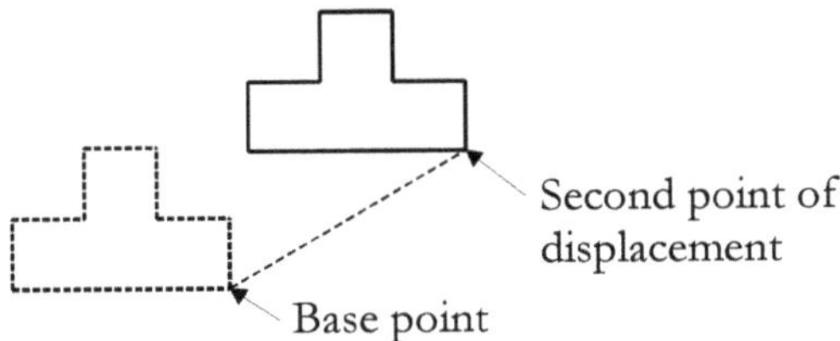

Figure 2.1 Moving objects using MOVE command

COPY Command

Modify toolbar: Copy
Pull-down menu: Modify → Copy
Command: COPY or CP ↵

The copy command is used to copy an existing object. This command is similar to the MOVE command, in the sense that it makes copies of the selected objects and places them at specified locations, but originals are left intact. You can also create multiple copies without invoking any special option (Figure 2.2).

```
Command: COPY ↵
Select objects: Select the object you want to copy
Select objects: Select another object or press ENTER
Current settings:  Copy mode = Multiple
Specify base point or [Displacement/mOde] <Displacement>:
                                   Specify a point on or near to the object
Specify second point or [Array] <use first point as displacement>:
                                   Select the location by specifying a point
Specify second point or [Array/Exit/Undo] <Exit>: ↵
```

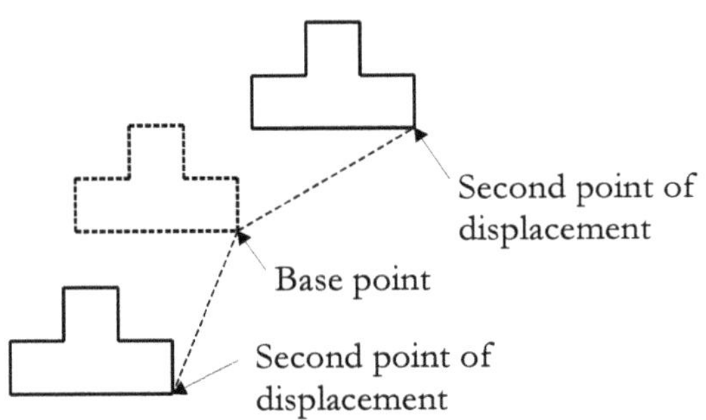

Figure 2.2 Making multiple copies using COPY command

ROTATE Command

Modify toolbar: Rotate
Pull-down menu: Modify → Rotate
Command: ROTATE or RO ↵

Sometimes, when making drawings, you may need to rotate an object or a group of object. You can accomplish this by using ROTATE command. How the object will be rotated is primarily depends on the defined base point as shown in Figure 2.3.

Command: **ROTATE ↵**
`Current positive angle in UCS: ANGDIR=counterclockwise ANGBASE=0`
`Select objects:` *Select the object for rotation*
`Select objects:` *Select another object or press ENTER*
`Specify base point:` *Specify a point on or near the object*
`Specify rotation angle or [Copy/Reference]:` *Enter the angle of rotation in degrees*

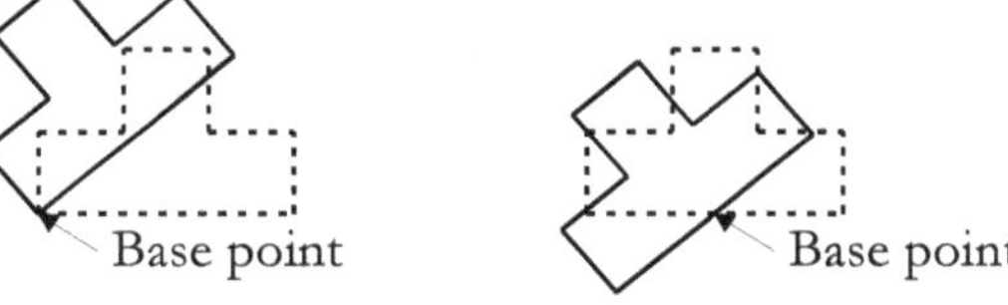

Figure 2.3 Rotation using ROTATE command

If you need to rotate objects with respect to a known angle, you can do this by using the Reference option.

Command: **ROTATE ↵**
`Current positive angle in UCS: ANGDIR=counterclockwise ANGBASE=0`
`Select objects:` *Select the object for rotation*
`Select objects:` *Select another or press ENTER*
`Specify base point:` *Specify a point on or near the object*
`Specify rotation angle or [Copy/Reference]:` **R ↵**
`Specify the reference angle <0>:` **10 ↵** (*You can also specify the angle by picking two points in the direction of your desired angle*)
`Specify the new angle or [Points]:` **100 ↵**

In this command, the actual rotation will be 90 degree. We can say, *rotation angle = new angle – reference angle.*

SCALE Command

Modify toolbar: Scale
Pull-down menu: Modify → Scale
Command: SCALE or SC ↵

Many times, you will need to change the size of the objects in a drawing. You can do it by using the SCALE command. This command enlarges or shrinks the selected objects in the same ratio for the *X* and *Y* dimensions about base point (Figure 2.4).

Command: **SCALE ↵**
Select objects: *Select object to be scaled*
Select objects: *Select another or press ENTER*
Specify base point: *Specify the base point of scaling*
Specify scale factor or [Copy/Reference]: *Enter the scale factor*

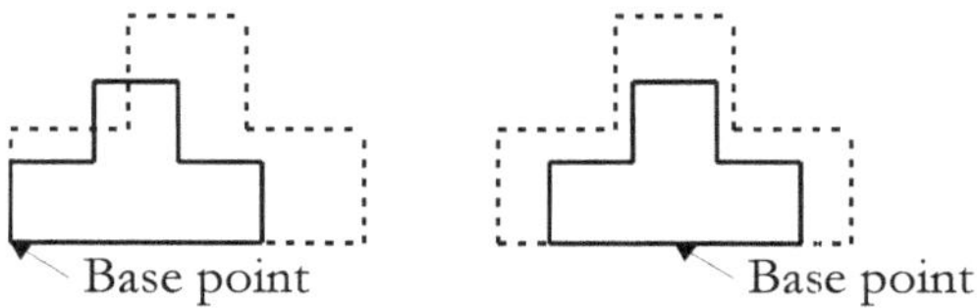

Figure 2.4 Using the SCALE command

Sometimes, it is time-consuming to calculate the relative scale factor. In such cases you can scale the object by specifying a desired size in relation to existing size. For example, you want to scale an object by scale factor 3/2. You can follow this procedure:

Command: **SCALE ↵**
Select objects: *Select object to be scaled*
Select objects: *Select another or press ENTER*
Specify base point: *Specify the base point of scaling*
Specify scale factor or [Copy/Reference]: **R ↵**
Specify reference length <1.0000>: **2 ↵**
Specify new length or [Points] <1.0000>: **3 ↵**

The *copy* option in the SCALE command enables you to retain the original in addition to the scaled object.

MIRROR Command

Modify toolbar: Mirror
Pull-down menu: Modify → Mirror
Command: MIRROR or MI ↵

The MIRROR command creates a mirrored copy of the selected objects. This command is helpful in drawing symmetrical figures (Figure 2.5). After invoking the command, AutoCAD will prompt to select the objects. Then AutoCAD prompts you to enter the beginning point and end point of mirror line (the imaginary line about which the objects will be reflected). Then AutoCAD prompts you to specify whether you want to retain the original figure or delete it.

`Command:` **MIRROR ↵**
`Select objects:` *Select the objects to be mirrored*
`Select objects:` *Select another object or press* ENTER
`Specify first point of mirror line:` *Specify the first endpoint*
`Specify second point of mirror line:` *Specify the second endpoint*
`Erase source objects? [Yes/No] <N>:` ↵ (*it retains the original figure*)

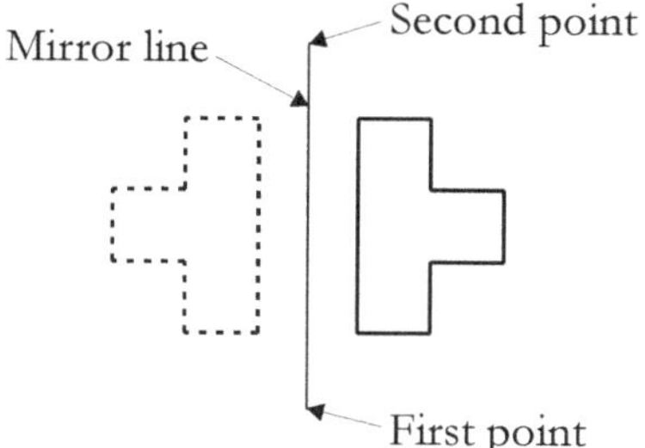

Figure 2.5 Reflecting objects using MIRROR command

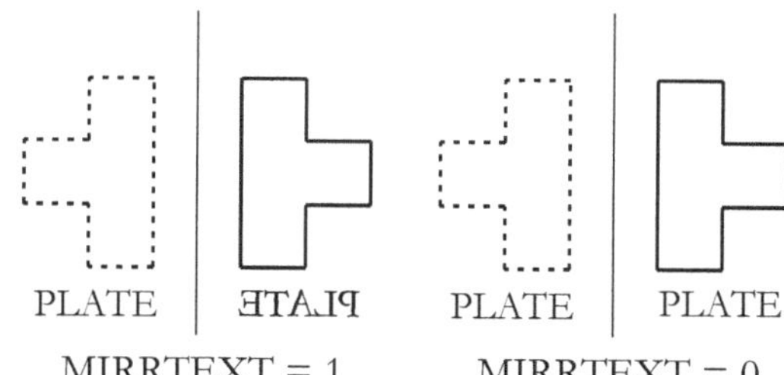

Figure 2.6 Using the MIRRTEXT variable to mirror the text

By default, the MIRROR command reverses all the objects, excluding texts. But you may want to reverse the text (written backward). In such cases you should use the system variable MIRRTEXT. This variable has the following two values.

1 = Text is reversed in relation to the original object (Figure 2.6).

0 = Resists the text from being reversed with respect to original object.

If you want a reversed text, set the MIRRTEXT variable to 1, before using the MIRROR command.

`Command:` **MIRRTEXT ↵**
`Enter new value for MIRRTEXT <0>:` **1↵**

OFFSET Command

Modify toolbar: Offset
Pull-down menu: Modify → Offset
Command: OFFSET or O ↵

You can draw parallel objects by using the OFFSET command. While offsetting an object, you can specify the offset distance and the side of offset, or you can specify a point through which you want to offset the selected object (Figure 2.7).

```
Command: OFFSET ↵
Current settings: Erase source=No Layer=Source OFFSETGAPTYPE=0
Specify offset distance or [Through/Erase/Layer] <Through>:
                                               Enter your desired distance
Select object to offset or [Exit/Undo] <Exit>: Select the object
Specify point on side to offset or [Exit/Multiple/Undo] <Exit>:
                                    Pick a point on your desired side of the object
Select object to offset or [Exit/Undo] <Exit>: Select another or
                                                          press ENTER
```

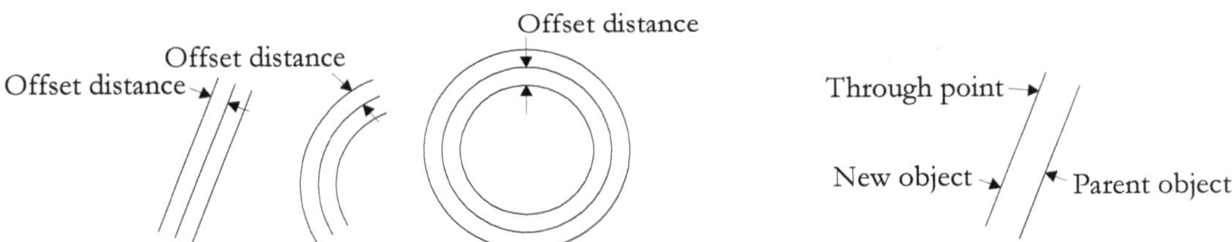

Figure 2.7 Using OFFSET command

The *through* option of OFFSET command is as follows:

```
Command: OFFSET ↵
Specify offset distance or [Through/Erase/Layer]: T ↵
Select object to offset or [Exit/Undo] <Exit>: Select the object
Specify through point or [Exit/Multiple/Undo] <Exit>: Specify the point
```

BREAK Command

Modify toolbar: Break
Pull-down menu: Modify → Break
Command: BREAK or BR ↵

The break command cuts existing objects into two parts or erases portions of an object. If you want to erase a portion of an object, following steps are to be followed (Figure 2.8):

```
Command: BREAK ↵
Select object: Select the object to be broken
Specify second break point or [First point]: F ↵
Specify first break point: Specify one point on the object
Specify second break point: Specify another point on the object
```

Notice that the portion of the object between specified two points has been erased. With this option, you can create a between two specified points on an object. You can use the default option also. In this case, at the time of selection, the selection point is taken as the first break point, and AutoCAD prompts for the second point.

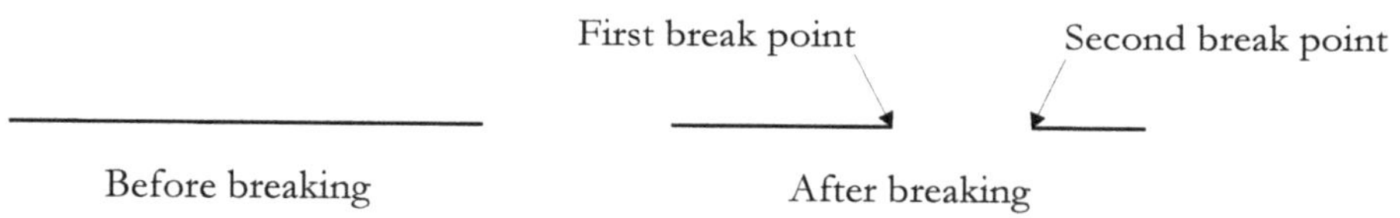

Figure 2.8 Using the BREAK command

If you want to break an object into two parts (instead of erasing a portion), follow the procedure:

```
Command: BREAK ↵
Select object: Select the object to be broken
Specify second break point or [First point]: @ ↵
```

In the above example, the selection point has been taken as the break point. If you want to specify a break point, other than the selection point, you have to follow this procedure:

```
Command: BREAK ↵
Select object: Select the object to be broken
Specify second break point or [First point]: F ↵
Specify first break point: Specify the breaking point on the object
Specify second break point: @ ↵
```

Now try to select the object and notice that the full object can not be selected at a single attempt, because of it has been broken into two parts.

TRIM Command

Modify toolbar: -/-- Trim
Pull-down menu: Modify → Trim
Command: TRIM or TR ↵

You may need to trim existing objects in a drawing. The TRIM command trims objects that extend beyond a required point of intersection. With this command you must select the cutting edges or boundaries first (Figure 2.9).

```
Command: TRIM ↵
Current settings: Projection=UCS, Edge=None
Select cutting edges.....
Select objects or <select all>: Select cutting edge
Select objects: Select second cutting edge or press ENTER
Select object to trim or shift-select to extend or
[Fence/Crossing/Project/Edge/eRase/Undo]: Select the first object to trim
Select object to trim or shift-select to extend or
[Fence/Crossing/Project/Edge/eRase/Undo]: Select second or press ENTER
```

You can use *edge* option whenever you want to extend objects that do not actually intersect the boundary edge, but would intersect its edge if the boundary edge was extended. Edge option also determines whether an object is trimmed at another object's implied edge or only to an object that intersects it in 3D space. Projection option specifies the projection that AutoCAD uses when trimming an object.

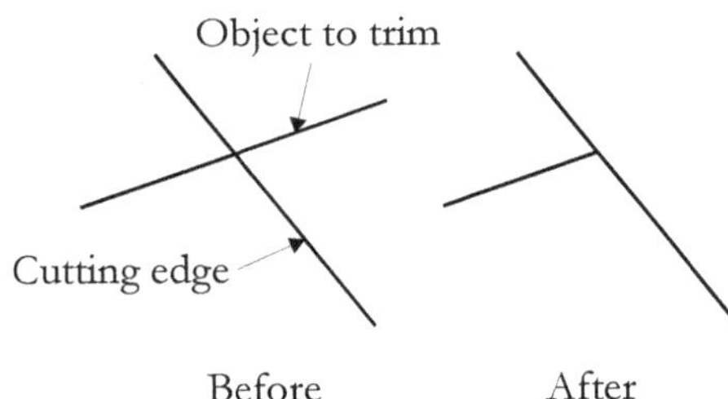

Figure 2.9 Using TRIM command

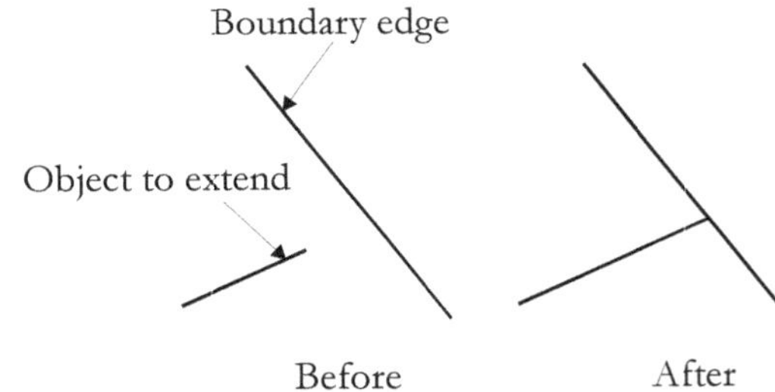

Figure 2.10 Using EXTEND command

EXTEND Command

Modify toolbar: --/ Extend
Pull-down menu: Modify → Extend
Command: EXTEND or EX ↵

If you want to extend an object up to another object, go through the EXTEND command. You are required to select the boundary edges first (Figure 2.10). The boundary edges are those objects up to which you want to extend.

```
Command: EXTEND ↵
Current settings: Projection=UCS, Edge=None
Select boundary edges...
Select objects or <select all>: Select the boundary edge
Select objects: Select another or press ENTER
Select object to extend or shift-select to trim or
[Fence/Crossing/Project/Edge/Undo]: Select the object you want to extend
Select object to extend or shift-select to trim or
[Fence/Crossing/Project/Edge/Undo]: Select another or press ENTER
```

> **Note:** You can extend object using TRIM command and trim an object using EXTEND command by holding down the Shift key while selecting the object.

LENGTHEN Command

Pull-down menu: Modify → Lengthen
Command: LENGTHEN or LEN ↵

Lengthen command can be used to extend or shorten a line or polyline, or an arc. This command has no effects on closed objects like circles, closed polylines etc.

```
Command: LENGTHEN ↵
Select an object or [DElta/Percent/Total/DYnamic]: Select the object
Current length: xx.xxxx When you select the object, AutoCAD will show the
                                                          current length
DElta/Percent/Total/DYnamic/<Select object>: T ↵
Specify total length or [Angle] <1.0000)>: Enter desired total length of
                                                          the object
Select an object to change or [Undo]: Select the object you want to change
Select an object to change or [Undo]: Select another or press ENTER
```

The *angle* option is applicable for arc or elliptical arc. The *percent* option is used to extend or shorten an object by defining the change as a percentage of the original length or angle. The *delta* option is used to increase or decrease the length or angle of an object by defining the delta distance or angle. Delta length is the length you want to extend or shorten. A positive value will increase the length and a negative value will decrease the length. The *dynamic* option allows you to dynamically change the length or angle of an object by specifying one of the endpoints.

ARRAY Command

Modify toolbar: Array
Pull-down menu: Modify → Array → Rectangular or Polar array
Command: ARRAY or AR ↵

In some instances, you may need to draw an object multiple times in a rectangular or circular (polar) arrangement. ARRAY command can help you in doing so. Following is an example for rectangular array (Figure 2.11):

Command: **ARRAY ↵**
Select objects: *Select the object*
Select objects: *Select another object or press ENTER*
Enter array type [Rectangular/PAth/POlar] <Rectangular>: ↵
Type = Rectangular Associative = Yes
Specify opposite corner for number of items or [Base point/Angle/Count] <Count>: ↵ *(press NULL ENTER)*
Enter number of rows or [Expression] <4>: *Enter number of rows*
Enter number of columns or [Expression] <4>: *Enter number of columns*
Specify opposite corner to space items or [Spacing] <Spacing>:
Enter a value to specify the spacing between rows and columns
Press Enter to accept or [ASsociative/Base point/Rows/Columns/Levels/eXit]<eXit>: ↵ *(press NULL ENTER)*

Associative option specifies whether to create items in the array as an associative array object, or as independent objects. *Rows* option edits the number and spacing of rows in the array. *Columns* option edits the number and spacing of columns.

Figure 2.11 Rectangular array

Following is the command sequence for circular (polar) array:

Command: **ARRAY ↵**
Select objects: *Select the object*
Select objects: *Select another object or press ENTER*
Enter array type [Rectangular/PAth/POlar] <Rectangular>: **PO ↵**
Type = Polar Associative = Yes
Specify center point of array or [Base point/Axis of rotation]:
Specify the center point

```
Enter number of items or [Angle between/Expression] <4>:
```
Enter the number of items you want in the array

```
Specify the angle to fill (+=ccw, -=cw) or [EXpression] <360>:
```
Enter the angle within which the objects will be arrayed

```
Press Enter to accept or [ASsociative/Base point/Items/Angle
between/Fill angle/ROWs/Levels/ROTate items/eXit]<eXit>: ↵
```

The *rotate items* option has very important role in polar array. It controls whether items are rotated as they are arrayed or not. Figure 2.12 and 2.13 explain the concept of rotation in an array.

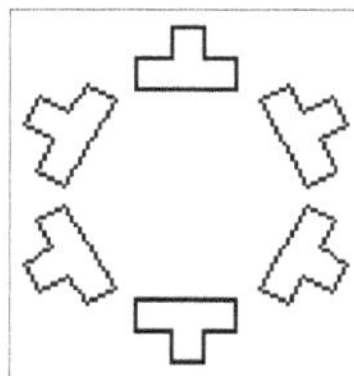

Figure 2.12 Rotated polar array

Figure 2.13 Non-rotated polar array

FILLET Command

Modify toolbar: Fillet
Pull-down menu: Modify → Fillet
Command: FILLET or F ↵

The FILLET command is used to create smooth round arcs to connect two objects (Figure 2.14). For changing the fillet radius, the prompt sequence is:

```
Command: FILLET ↵
Current settings: Mode = TRIM, Radius = 0.0000
Select first object or [Undo/Polyline/Radius/Trim/Multiple]: R ↵
Specify fillet radius <0.0000>: Enter your desired fillet radius
Select first object or [Undo/Polyline/Radius/Trim/Multiple]:
                                                 Select first object
Select second object or shift-select to apply corner or [Radius]:
                                                   Select 2nd object
```

If you want to fillet a closed pline, invoke *polyline* option and select the polyline. *Trim* option controls TRIMMODE. If this option is set to *trim*, the selected objects are either trimmed or extended to the fillet arc endpoints. If set to *no trim* they are left intact. *Multiple* option lets you to perform multiple fillets at a time, otherwise by default it offers one fillet operation only.

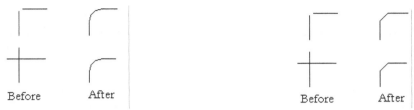

Figure 2.14 Using FILLET command   **Figure 2.15** Using CHAMFER command

CHAMFER Command

Modify toolbar: Chamfer
Pull-down menu: Modify → Chamfer
Command: CHAMFER or CHA ↵

In the engineering drawing, the CHAMFER is defined as the taper provided on a surface. Sometimes the chamfer is used to avoid a sharp corner. In AutoCAD, a chamfer is any angled corner of a drawing (Figure 2.15). Following is command sequence:

```
Command: CHAMFER ↵
(TRIM mode) Current chamfer Dist1 = 0.0000, Dist2 = 0.0000
Select first line or [Undo/Polyline/Distance/Angle/Trim/
mEthod/Multiple]: D ↵
Specify first chamfer distance <0.0000>: Enter the distance
Specify second chamfer distance <2.0000>: Enter the distance
Select first line or [Undo/Polyline/Distance/Angle/Trim/
mEthod/Multiple]: Select the first line
Select second line or shift-select to apply corner or
[Distance/Angle/Method]: Select the second line
```

If you want to chamfer a closed pline, invoke *polyline* option and select the polyline to be chamfered. Trim option controls TRIMMODE. If this option is set to *trim*, the selected objects are either trimmed or extended to the chamfer line endpoints. If set to *no trim* they are left intact. *Multiple* option lets you do multiple chamfers at a time, otherwise by default it offers one chamfer operation only.

DDEDIT Command

Text toolbar: Edit...

Pull-down menu: Modify → Object → Text → Edit....
Command: DDEDIT or ED ↵

If you want to change a text of your drawing, go through the DDEDIT command.

Command: **DDEDIT ↵**
Select an annotation object or [Undo]: *Select the text and edit it*
Select an annotation object or [Undo]: ↵

If the selected text is a multiline text (MTEXT), the Multiline Text Editor dialog box will appear (Figure 1.19, Chapter 1). You have learned about this editor in chapter 1. If the text is dynamic text (DTEXT), the text will be highlighted and invoked in editable mode (Figure 2.16). Edit your text and press ENTER.

ALL DIMENSIONS ARE IN MM

Figure 2.16 Dynamic text in editable mode

STRETCH Command

Modify toolbar: Stretch
Pull-down menu: Modify → Stretch
Command: STRETCH or S ↵

This command can be used to stretch objects, altering selected portions of the objects. With the help of this command you can lengthen objects, shorten them, and alter their shapes. You must have to use crossing-window or crossing-polygon selection (discussed at the end of this chapter) to specify the objects to stretch. The prompt sequence is:

Command: **STRETCH ↵**
Select objects to stretch by crossing-window or crossing-polygon...
Select objects: *Specify a point to select objects*
Specify opposite corner: *Specify a point so that it creates a crossing window/polygon*
Select objects: ↵
Specify base point or [Displacement] <Displacement>: *Specify base point*
Specify second point or <use first point as displacement>: *Specify the displacement point*

HATCHEDIT Command

Modify II toolbar: Edit Hatch
Pull-down menu: Modify → Object → Hatch...
Command: HATCHEDIT or HE ↵

With this command you can edit a hatch previously drawn by you.

```
Command: HATCHEDIT ↵
Select hatch object: Select the hatch you want to edit
```

Hatchedit dialog box will appear. This dialog box is same as Boundary Hatch dialog box (Figure 1.20, Chapter 1). Change as you desire and click on OK button.

PEDIT Command

Modify II toolbar: Edit Polyline
Pull-down menu: Modify → Object → Polyline
Command: PEDIT or PE ↵

You can use the PEDIT command to edit any type of Polyline. The prompt sequence:

```
Command: PEDIT ↵
Select polyline or [Multiple]: Select the polyline to be edited
Enter an option [Close/Join/Width/Edit vertex/Fit/Spline/Decurve/
Ltype gen/Reverse/Undo]: Select a specific option for your desired change
```

PEDIT command offers the following options to edit polylines:

M (Multiple): This option enables you to select more than one object.

C (Close): This option creates the closing segment of the polyline, connecting the last segment with the first. AutoCAD considers the polyline open unless you close it by using the Close option.

O (Open): If the selected polyline is closed, the close option is replaced by the open option. Entering O, for open, removes the closing (last) segment.

J (Join): This option adds lines, arcs, or polylines to the end of an open polyline and removes the curve fitting from a curve-fit polyline. For an object to join the polyline, their endpoints must touch.

W (Width): This option is useful to change the line width of all segments of the selected polyline.

E (Edit vertex): This option lets you select a vertex of a polyline and perform different editing operations on the vertex and the segments following it.

F (Fit): This option creates a smooth curve consisting of arcs joining each pair of vertices. The curve passes through all vertices of the polyline and uses any tangent direction you specify.

S (Spline): This option also smoothes the corners of a straight segment polyline, as does the *fit* option, but the curve passes through only the first and the last vertices, (except in the case of a closed polyline).

D (Decurve): This option straightens the curves generated after using the *fit/spline* option or drawn by *arc* option.

L (Ltypegen): This option generates the linetype in a continuous pattern through the vertices of the polyline. When this option is turned off, AutoCAD generates the linetype starting and ending with a dash at each vertex.

R (Reverse): This option reverses the order of vertices of the polyline. Use this option to reverse the direction of objects that use linetypes with included text. For example, depending on the direction in which a polyline was created, the text in the linetype might be displayed upside down.

MEASURE Command

Pull-down menu: Draw → Point → Measure
Command: MEASURE or ME ↵

You may need to segment an object at fixed distances without actually dividing it. You can use the measure command to do so. Before invoking the MEASURE command you should set the point style as marker.

`Command:` **MEASURE ↵**
`Select object to measure:` *Select the object to be measured*
`Specify length of segment or [Block]:` *Enter measuring length*

The *block* option places blocks at a specified interval along the selected object.

`Enter name of block to insert:` *Enter the name of a block currently defined in the drawing*

`Align block with object? [Yes/No] <Y>:` *Enter Y or N*

If you enter Y, the block is rotated about its insertion point so that its horizontal lines are aligned with, and drawn tangent to, the object being measured. If you enter N, the block is always inserted with a 0 rotation angle.

`Specify length of segment:` *Enter measuring length*

After you specify the segment length, AutoCAD inserts the block at the specified interval. If the block has variable attributes, these attributes are not included.

DIVIDE Command

Pull-down menu: Draw → Point → Divide
Command: DIVIDE or DIV ↵

DIVIDE command marks off a specified number of equal lengths on a selected object by placing point objects or blocks along the length or perimeter of the object. Objects that you can divide include arcs, circles, ellipses and elliptical arcs, polylines, and splines. If you are not using block, you should set a point style before invoking the DIVIDE command.

`Command:` **DIVIDE ↵**
`Select object to divide:` *Select the object*
`Enter the number of segments or [Block]:` *Enter value from 2 to 32,767*

The *block* option is similar to the block option of MEASURE command.

LINETYPE Command

Pull-down menu: Format → Linetype...
Command: LINETYPE or LT ↵

If you want to change the line-type of an object you have to load that type of line at first. As an example if you want to draw a center line, you have to load the center linetype. After invoking the command, Linetype Manager dialog box (Figure 2.17) will appear on the screen. Click on Load button, Load or Reload Linetypes dialog box (Figure 2.18) will appear. Select the linetype you want to load and click on OK.

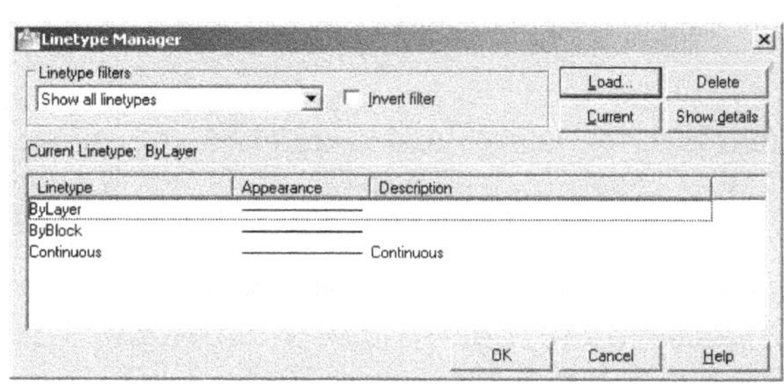

Figure 2.17 Linetype Manager dialog box

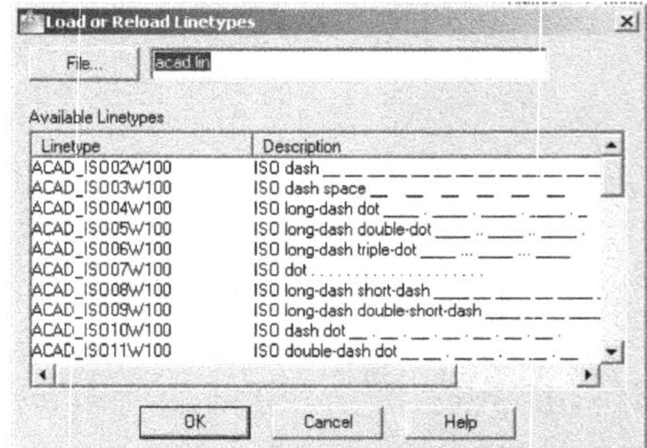

Figure 2.18 Load or Reload Linetypes dialog box

Details of Linetype Manager dialog box

List of Linetypes: Displays the loaded linetypes according to the option specified in Linetype Filters. To select all or clear all linetypes quickly, right-click in the linetype list to display the shortcut menu.

Load: Displays the Load or Reload Linetypes dialog box, in which you can load into the drawing selected linetypes from the acad.lin file and add them to the linetype list.

Current: Sets the selected linetype to be the current linetype. Setting the current linetype to BYLAYER means an object assumes the linetype that is assigned to a particular layer. Setting the linetype to BYBLOCK means an object assumes the CONTINUOUS linetype until it is grouped into a block. Whenever the block is inserted, all objects inherit the block's linetype. The CELTYPE system variable stores the linetype name.

Delete: Deletes the selected linetypes from the list. You can delete unreferenced linetypes only. Default referenced linetypes include BYLAYER, BYBLOCK, and CONTINUOUS. Be careful about deleting linetypes if you are working on a drawing in a shared project or one based on a set of layering standards. The deleted linetype definition remains stored in the acad.lin or acadiso.lin file and can be reloaded.

Show Details or Hide Details: Controls whether the Details section of the Linetype Manager is displayed.

LTSCALE Command

Command: LTSCALE or LTS ↵

You can change the linetype scale globally by this command. After invoking the command this prompt will appear

`Enter new linetype scale factor <1.0000>:` *Enter your desired factor*

Example: Linetype scale 1: ______ __ ______ __ ______
Linetype scale 0.5: ___ _ ___ _ ___ _ ___ _ ___ _ ___

DDCHPROP or CH Command

Standard toolbar: Properties

Pull-down menu: Modify → Properties

Command: DDCHPROP or CH or PROPERTIES or PR ↵
(or press *Ctrl+1* from keyboard)

This command may also be accessed from shortcut menu. Select the objects whose properties you want to view or modify, right-click in the drawing area, and choose Properties. Alternatively, you can double-click on the objects to display the Properties palette.

This command displays the properties of the selected object or set of objects. You can specify a new value to modify any property that can be changed. When more than one object is selected, the Properties palette displays only those properties common to all objects in the selection set. When no objects are selected, the Properties palette displays only the general properties of the current layer, the name of the plot style table attached to the layer, the view properties, and information about the UCS.

After invoking the command the Properties palette dialog box (Figure 2.19) will appear in which you can view or change the properties.

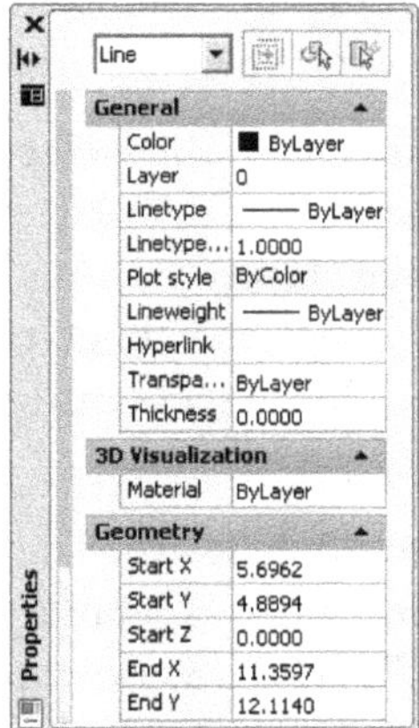

Figure 2.19 Properties palette

Object Selection Options

There are several selection options in AutoCAD, which you have not tried yet. When AutoCAD prompts '`Select object:` ', till now you select your desired objects individually one by one. There are several other options as explained below.

All option selects all the objects in a drawing except those in frozen or locked layers (layers will be discussed later).

Window option lets you select objects by enclosing them in a rectangular window. A window is automatically produced when a point is picked and no object is found. A window is produced when the two window corner is picked from left to right.

Crossing window option is similar to window, but selects everything that encloses or crosses through the window you defined. A crossing window is produced when the two corners are picked from right to left.

Last option selects the last object you have drawn.

Previous option selects the objects that were selected by you in previous command.

Single option forces the command to select only a single object.

Fence option selects objects that are crossed over by a temporary line called a fence. This option is like crossing out the objects you want to select.

Remove option is used to remove an object from the selection set.

Add option can be used for coming back to the Add mode from the Remove mode.

Undo option removes the most recently selected object from the selection set.

Object Selection using FILTER

Command: FILTER or FI ↵ ('FILTER or 'FI for transparent use)

Filter command displays the Object Selection Filters dialog box (Figure 2.20). You can select objects by using the filter properties that compose the current filter. The current filter is the filter that you select in Current in the Named Filters area. You can use the logical operators, such as AND, OR and NOT. For example, the following filter selects all circles except with a radius greater than or equal to 1.0:

```
Object =Circle
**Begin NOT
Circle Radius >= 1.00
**End NOT
```

In the filter parameters, you can use relative operators such as < (less than) or > (greater than). For example, the following filter selects all circles with center point coordinates greater than or equal to 1,1,0 and radii greater than or equal to 1:

```
Object = Circle
Circle Center X >= 1.0000 Y >= 1.0000 Z >= 0.0000
Circle Radius >= 1.0000
```

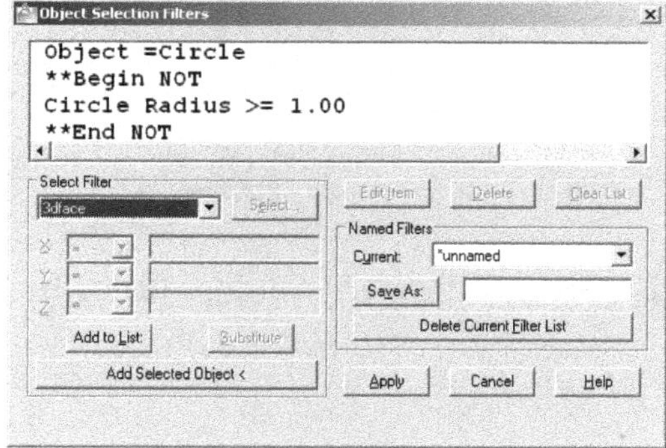

Figure 2.20: Object Selection Filters dialog box

MATCHPROP Command

Standard toolbar: Match Properties

Pull-down menu: Modify → Match Properties

Command: MATCHPROP or MA ↵

This command applies the properties of a selected object to other objects.

`Command:` **MATCHPROP ↵**

`Select source object:` *Select the object whose properties you want to copy*

`Select destination object(s) or [Settings]:` *Enter S for settings or select one or more objects to copy properties to*

`Select destination object(s) or [Settings]:` ↵

Destination object(s) specifies the objects to which you want to impose the properties of the source object. You can continue selecting destination objects, or press ENTER to apply the properties and end the command.

The *settings* option displays the Property Settings dialog box, in which you can control which object-properties to copy to the destination objects. By default, AutoCAD selects all object properties in the Property Settings dialog box (Figure 2.21) for copying. You can select specified properties to be copied with MATCHPROP. Select one or more of these settings from the Property Settings dialog box.

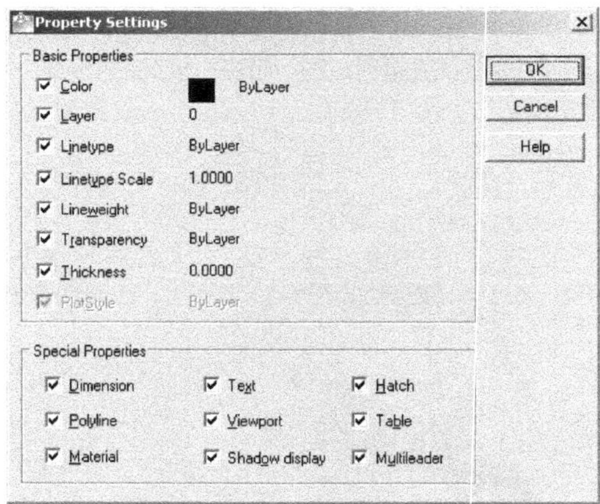

Figure 2.21 Property Settings dialog box

3

Display Techniques and Dimensioning

Introduction

In this chapter you will learn about some commands which will help you for better viewing, setting up the drawing area, setting the scale and unit, drawing in different layers, dimensioning, setting dimension styles, and printing/plotting.

ZOOM Command

ZOOM Command

Standard tools:

Pull-down menu: View → Zoom

Command: ZOOM or Z ('ZOOM or 'Z for transparent use) ↵

Shortcut menu: With no objects selected right-click in the drawing area and choose Zoom to zoom in real time.

Often you may need to magnify your drawing for better viewing. You can magnify or reduce the view of the drawing on the screen using ZOOM command. ZOOM command magnifies the object, but it does not affect the actual size of the drawing in print.

```
Command: ZOOM ↵
Specify corner of window, enter a scale factor (nX or nXP),
Or [All/Center/Dynamic/Extents/Previous/Scale/Window/
Object] <real time>: Enter an option
```

Some frequently used zooming options

Extents option: As the name indicates, this option lets you zoom to the extent of the drawing. Drawing extent means the imaginary smallest rectangular area over the entire drawing.

All option: This option of ZOOM command displays the drawing limits or extents, whichever is greater.

Window option: It is the most useful option of ZOOM command. It lets you specify the area you want to zoom in by specifying two opposite corners of a rectangular window.

Previous option: If you have zoomed your drawing and now want to go back to the previous view this option is useful. Previous option can restore previous 10 views.

Center: This option zooms to display a window area defined by a center point and a magnification value or height. A smaller value for the height increases the magnification. A larger value decreases the magnification.

Dynamic: This option zooms to display the generated portion of the drawing with a view box. The view box represents your viewport, which you can shrink or enlarge and move around the drawing. Positioning and sizing the view box pans or zooms to fill the viewport with the image inside the view box.

Scale: This option zooms the display at a specified scale factor.

Object: This option zooms to display one or more selected objects as large as possible and in the center of the view.

Real time: Using the pointing device, you can zoom interactively to a logical extent. The cursor changes to a magnifying glass with plus (+) and minus (–) signs. This option of ZOOM uses half of the window height to move to a zoom factor of 100%. Holding down the pick button at the midpoint of the window and moving vertically to the top of the window zooms into 100%. Conversely, holding the pick button down at the midpoint of the window and moving vertically to the bottom of the window zooms out by 100%.

PAN Command

Standard toolbar: Pan
Pull-down menu: View → Pan
Command: PAN or P ('PAN or 'P for transparent use) ↵
Shortcut menu: With no objects selected right-click in the drawing area and choose Pan.

You may want to view or draw on a particular area outside of the current viewport. You can do this by using the PAN command. Pan allows you to view different portions of your drawing by moving the cursor, displayed as an open hand, in any direction. Once panning is done, press Esc or Enter to exit, or right-click to activate pop-up menu, and then click on Exit.

LIMITS Command

Pull-down menu: Format → Drawing Limits
Command: LIMITS ↵

When you start AutoCAD, the default limits are 12, 9. That means your drawing area is 12 units along *X*-axis and 9 units along *Y*-axis. You may use the LIMITS command to change your drawing area.

`Command:` **LIMITS ↵**
`Reset Model space limits:`
`Specify lower left corner or [ON/OFF] <0.0000,0.0000>:` *Press ENTER to accept the default value or change it*
`Specify upper right corner <12.0000,9.0000>:` *Press ENTER to accept default value or change it*

After changing the drawing limits, you must have to go through the ZOOM command with *all* option. Otherwise the changed limits will not be displayed on your screen.

LAYER Command

Layers toolbar: Layer Properties Manager
Pull-down menu: Format → Layer...
Command: LAYER or LA ↵

Layers are the equivalent of the overlays used in paper-based drafting. They are the primary organizational tool in AutoCAD. You can use them to group information by function and to enforce linetype, color, and other standards. By creating layers, you can associate similar types of objects by assigning them to the same layer (Figure 3.1). For example, you can put construction lines, text, dimensions, and title blocks on separate layers. When you begin a new drawing, AutoCAD creates a special layer named 0. By default, layer 0 is assigned color number 7, the continuous linetype, a lineweight of default (the default setting is .01 inch or .25 mm), and the *normal* plot style. Layer 0 cannot be deleted or renamed.

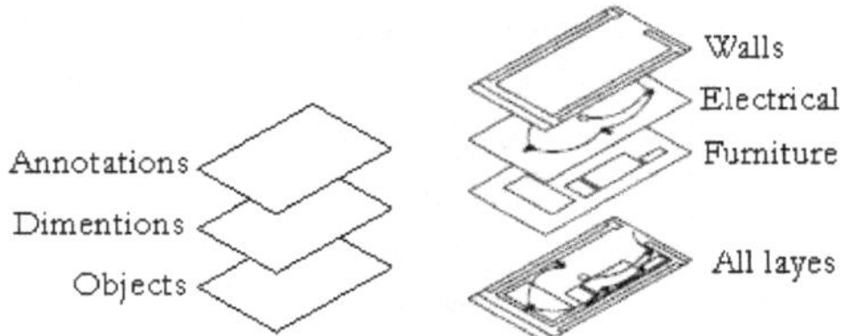

Figure 3.1 Sample Layers

The LAYER command displays the Layer Properties Manager dialog box (Figure 3.2). This dialog box makes a layer current, adds new layers, deletes existing layers, and renames layers. You can assign properties to layers, turn layers on and off, freeze and thaw globally or by viewport, lock and unlock, set plot styles, and turn plotting on and off. You can filter the layer names displayed in the Layer Properties Manager, and you can save and restore layer states and properties settings.

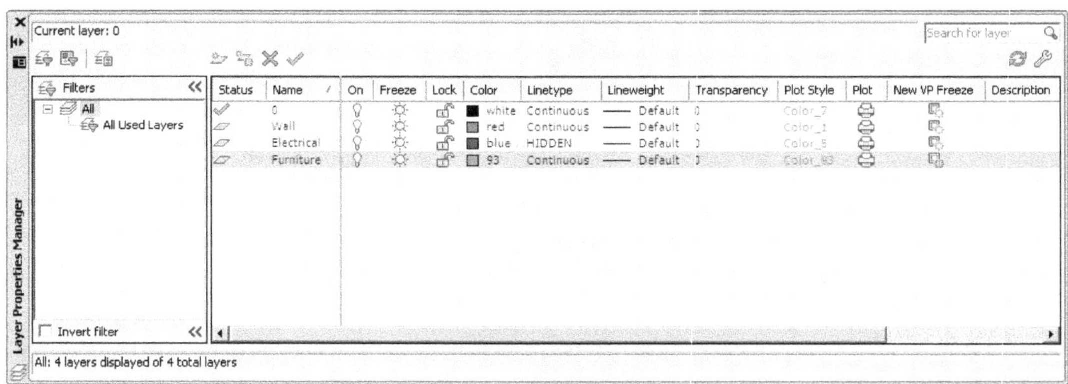

Figure 3.2 Layer Properties Manager dialog box

Following are the important options to change the status/properties of a layer

Lock option: You can not edit the objects of locked layer, but you can see and print those objects.

Unlock option: It unlocks selected locked layers by permitting editing on those layers.

Freeze option: This option freezes layers by making them invisible and excluding them from regeneration and plotting.

Thaw option: It thaws frozen layers by making them visible and available for regeneration and plotting.

OFF option: This option is similar to Freeze. You can not see or print the objects. But you can select the objects by *all* option of object selection procedure. Further, regeneration occurs.

> **Note:** Regeneration of drawing means converting the floating points to integer for display purpose; because the display device can not understand any floating value. AutoCAD regenerates automatically when required. But you can forcefully regenerate your drawing by the command REGEN (or RE in short).

ON option: This option makes selected layers visible and available for plotting.

Linetype option: It changes the linetype associated with a layer.

Color option: The color option changes the color associated with a layer.

Plot/Don't Plot option: It controls whether the selected layers will be plotted. If you turn off plotting for a layer, the objects on that layer are still displayed but will not appear on the printout. Turning off plotting for a layer affects only visible layers in the drawing (layers which are on and thawed). Turning off plotting for layers containing reference information such as construction lines can be useful.

Lineweight option: This option changes the lineweight associated with the selected layers. Clicking any lineweight name displays the Lineweight dialog box.

You can change your current layer and the status/properties of layers from Layers toolbar (Figure 3.3). You can also handle several options of LAYER command from this toolbar.

Figure 3.3 Layers toolbar

ID Command

Inquiry toolbar: Locate Point
Pull-down menu: Tools → Inquiry → ID Point
Command: ID ↵

If you want to know the coordinate of a point you have to go through this command. After invoking the command AutoCAD will ask for the point and after defining the point it will display the coordinate of the point.

DIST Command

Inquiry toolbar: Distance
Pull-down menu: Tools → Inquiry → Distance
Command: DIST ↵

By this command you can measure the distance between two points. The command sequence is:

```
Command: DIST ↵
Specify first point: Specify a point from where distance measurement will start
Specify second point or [Multiple points]: Specify the point where
                                          the distance measurement will end
```

AutoCAD responds with the following information:

```
Distanc=3.2159, Angle in XY Plane=28, Angle from XY Plane=0
Delta X = 2.8344,  Delta Y = 1.5192,   Delta Z = 0.0000
```

LIST Command

Inquiry toolbar: List
Pull-down menu: Tools → Inquiry → List
Command: LIST ↵

By this command you can get information about one or more objects. The command asks for selecting the objects for which you want the information. Following is the example for a line:

```
            LINE      Layer: 0
                      Space: Model space
             Color: BYLAYER    Linetype: CENTER
             Handle = 55
        from point, X= 4.2747   Y= 2.0831   Z= 0.0000
          to point, X= 6.6022   Y= 3.4182   Z= 0.0000
            Length = 2.6832,   Angle in XY Plane = 30
      Delta X = 2.3275,  Delta Y = 1.3350,  Delta Z = 0.0000
```

AREA Command

Inquiry toolbar: Area
Pull-down menu: Tools → Inquiry → Area
Command: AREA ↵

If you want to know the area and perimeter of a closed object, you can go through the AREA command. You can also calculate the area and perimeter by giving some points on your drawing. The command sequence is:

```
Command: AREA ↵
Specify first corner point or [Object/Add area/Subtract area]
<Object>: Specify a point

Specify next point or [Arc/Length/Undo]: Specify another point
Specify next point or [Arc/Length/Undo]: Specify another point
Specify next point or [Arc/Length/Undo/Total]<Total>: Specify another point
Specify next point or [Arc/Length/Undo/Total] <Total>: ↵
```

Now AutoCAD will show the area and perimeter. You can know the area of a closed object by invoking the *object* option at the first line of prompt sequence.

MVSETUP Command

You can draw your drawing in any size by changing the drawing limits. But if you want to setup the sheet size, scale factor, type of unit and also want a border line around the drawing area, it is recommended to go through the MVSETUP command. When you enter MVSETUP at the command line, the prompts displayed depend on whether you are on the *model tab* (model space) or on a *layout tab* (paper space).

On the model tab, you can set the units type, drawing scale factor, and paper size. Using the settings you provide, AutoCAD draws a rectangular border at the drawing limits.

On a layout tab, you can insert one of several predefined title blocks into the drawing and create a set of layout viewports within the title block. You can specify a global scale as the ratio between the scale of the title block in the layout and the drawing on the model tab. Let us first consider the model tab (model space).

```
Command: MVSETUP ↵
Initializing...
Enable paper space? (No/<Yes>): N ↵
Enter units type (Scientific/Decimal/Engineering/
Architectural/Metric): M ↵

Metric Scales
=================
 (5000) 1:5000
 (2000) 1:2000
```

```
(1000) 1:1000
(500)  1:500
(200)  1:200
(100)  1:100
(75)   1:75
(50)   1:50
(20)   1: 20
(10)   1:10
(5)    1:5
(1)    FULL
```

(These scale factors will be shown by the AutoCAD. But it is not essential to use these scale factors. You may use your own choice.)

`Enter the scale factor:` *Enter the scale factor for the drawing*
`Enter the paper width:` *Enter the paper width*
`Enter the paper height:` *Enter the paper height*

AutoCAD supported unit types are as follows

Formats	*Examples*
1. Scientific	1.55E+01
2. Decimal	15.50
3. Engineering	1'-3.50"
4. Architectural	1'-3 1/2"
5. Fractional	15 1/2

Standard sheet sizes

Sheet	*in mm*	*in inch*
A0	1189 × 841	46.8 × 33.1
A1	841 × 595	33.1 × 23.4
A2	594 × 420	23.4 × 16.5
A3	420 × 297	16.5 × 11.7
A4	297 × 210	11.3 × 8.3

Remember, the aforementioned sheet sizes are actual dimensions of the sheet (trimmed size). A printer or plotter can not print in the entire paper. Plotted area of a sheet is little less than the actual sheet size. It is recommended to subtract 20 mm from each dimension of the sheet. For examples, A4 sheet should be considered as 277 mm × 190 mm instead of 297 mm × 210 mm. Similarly, A3 sheet should be considered as 400 mm × 277 mm instead of 420 mm × 297 mm. It is advised to use reduced dimensions of sheet size in MVSETUP command.

If you want to work on layout tab (paper space) you have to turn off the TILEMODE system varriable. When you turn off the TILEMODE system variable, the Page Setup dialog box appears on the screen. Now setup the page size for the new layout. To easily specify all layout page settings and prepare your drawing for plotting, you can use the Page Setup dialog box from File menu also, which is automatically displayed when you select a layout in a new drawing session.

VPORTS Command

Layouts toolbar: Vports
Pull-down menu: View → Viewports
Command line: VPORTS ↵

This command creates multiple viewports in a drawing. The number and layout of active viewports and their associated settings are called viewport configurations. This command determines the viewport configuration for model space and paper space (layout) environments. In model space (the model tab), you can create multiple model viewport configurations. In paper space (a layout tab), you can create multiple layout viewport configurations.

Dimensioning

To make your designs more informative and practical, the drawing must convey more than just the graphic picture of the product. To manufacture an object, the drawing must contain size descriptions such as width, height, angle, radius, diameter etc. All these data are added to the drawing with the help of dimensioning.

Horizontal and Vertical Dimensioning

Dimension toolbar: Linear
Pull-down menu: Dimension → Linear
Command: DIMLIN or DIMLINEAR ↵

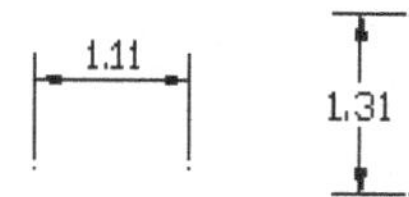

Figure 3.4 Linear (horizontal and vertical) dimensions

The prompt sequence is as follows:

```
Command: DIMLIN ↵
Specify first extension line origin or <select object>: Select first origin
Specify second extension line origin: Select second origin
Specify dimension line location or [Mtext/Text/Angle/ Horizontal/
Vertical/Rotated]: Pick a point through which the dimension line will pass
Dimension text = <default>
```

Mtext and Text option: AutoCAD writes (draws) the measured actual associative dimension text. You can change the default dimension text with these options.

Angle option: This option lets you change the angle of the dimension text.

Horizontal option: This option lets you create a horizontal dimension regardless of where you specify the dimension location.

Vertical option: This option lets you create a vertical dimension regardless of where you specify the dimension location.

Rotated option: This option lets you create a dimension that is rotated at a specified angle.

Aligned Dimensioning

Dimension toolbar: Aligned
Pull-down menu: Dimension → Aligned
Command: DIMALI or DIMALIGNED ↵

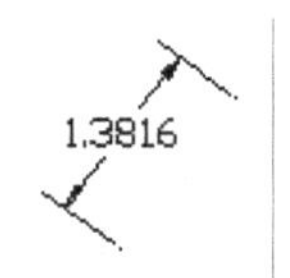

Figure 3.5 Aligned dimension

Prompt sequence:

```
Command: DIMALI ↵
Specify first extension line origin or <select object>: Select first origin
Specify second extension line origin: Select second origin
Specify dimension line location or [Mtext/Text/Angle]: Pick a point
Dimension text = <default>
```

Radius Dimensioning

Dimension toolbar: Radius
Pull-down menu: Dimension → Radius
Command: DIMRAD or DIMRADIUS ↵

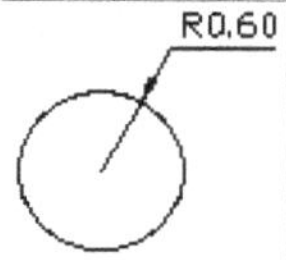

Figure 3.6 Radius dimension

Prompt sequence:

```
Command: DIMRAD ↵
Select arc or circle: Select the arc or circle
Dimension text = <default>
Specify dimension line location or [Mtext/Text/Angle]: Pick a point
```

Diameter Dimensioning

Dimension toolbar: Diameter
Pull-down menu: Dimension → Diameter
Command: DIMDIA or DIMDIAMETER ↵

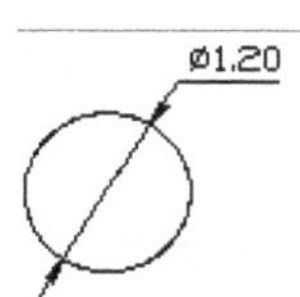

Figure 3.7 Diameter dimension

Prompt sequence:

```
Command: DIMDIA ↵
Select arc or circle: Select the arc or circle
Dimension text = <default>
Specify dimension line location or [Mtext/Text/Angle]: Pick a point
```

Angular Dimensioning

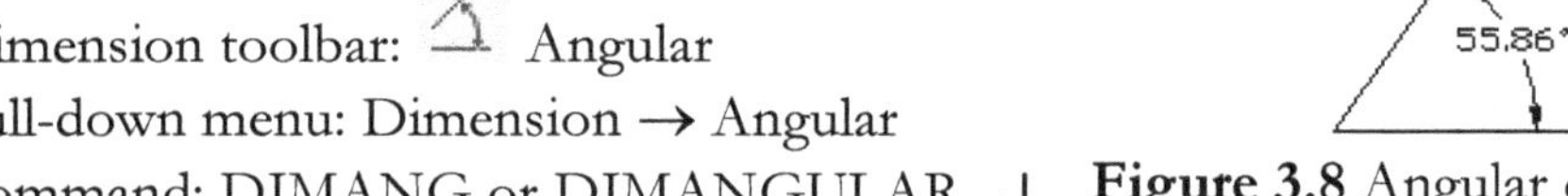

Dimension toolbar: Angular
Pull-down menu: Dimension → Angular
Command: DIMANG or DIMANGULAR ↵

Figure 3.8 Angular dimension

Prompt sequence:

Command: **DIMANG ↵**
Select arc, circle, line, or <specify vertex>: *Select first line*
Select second line: *Select second line*
Specify dimension arc line location or [Mtext/Text/Angle]: *Pick a point*
Dimension text = <default>

Leader Dimensioning

Command: LE ↵

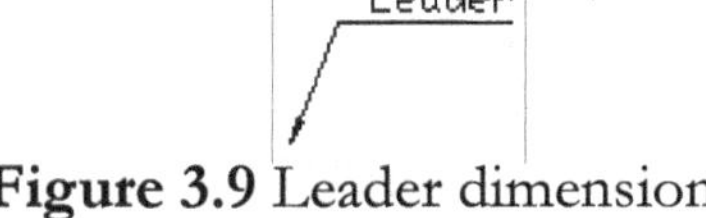

Figure 3.9 Leader dimension

Prompt sequence:

Command: **LE ↵**
Specify first leader point, or [Settings] <Settings>: *Pick starting point*
Specify next point: *Pick next point*
Specify next point: *Pick next point*
Specify next point: *Pick next point*
Specify next point: ↵
Specify text width <0.0000>: ↵
Enter first line of annotation text <Mtext>: *Type the dimension text*
Enter next line of annotation text: *Type next line of dimension text or press* ENTER
Enter next line of annotation text: *Type next line of dimension text or press* ENTER

Baseline Dimensioning

Dimension toolbar: Baseline
Pull-down menu: Dimension → Baseline
Command: DIMBASE ↵

This command creates a linear, angular, or ordinate dimension from the baseline of the previous dimension or a selected

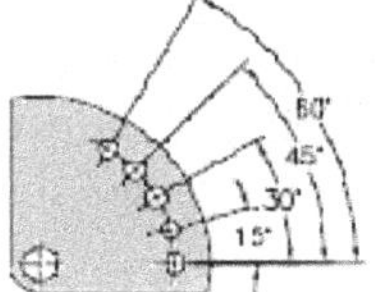

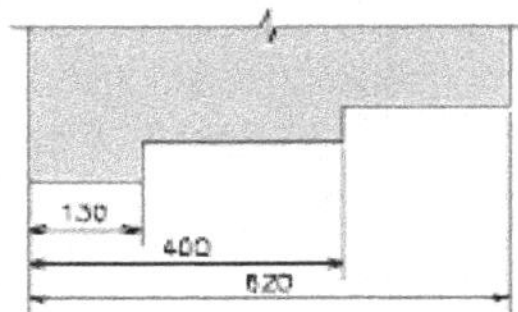

Figure 3.10 Angular and linear Baseline dimensioning

dimension. DIMBASELINE creates a series of related dimensions measured from the same baseline (Figure 3.10). AutoCAD uses a baseline increment value to offset each new dimension line and to avoid overlaying the previous dimension line. The default spacing between baseline dimensions can be set from the Dimension Style Manager, Lines tab, Baseline Spacing.

Prompt sequence:

Command: **DIMBASE ↵**

Specify a second extension line origin or [Undo/Select] <Select>: ↵
Select base dimension: *Select the base dimension*
Specify a second extension line origin or [Undo/Select] <Select>:
Dimension text = <Default>
Specify a second extension line origin or [Undo/Select] <Select>:
Dimension text = <Default>
Specify a second extension line origin or [Undo/Select] <Select>: ↵

Continuous Dimensioning

Dimension toolbar: Continue
Pull-down menu: Dimension → Continue
Command: DIMCONT ↵

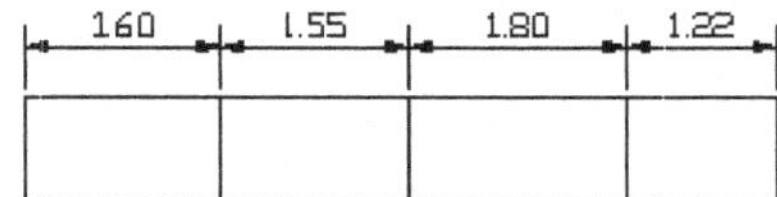

Figure 3.11 Continuous dimension

This command creates a linear, angular, or ordinate dimension from the second extension line of the previous dimension or a selected dimension (Figure 3.11). DIMCONTINUE draws a series of related dimensions, such as several shorter dimensions which add up to the total measurement. Continued dimensioning is also known as chain dimensioning. When you create linear continued dimensions, the first extension line is suppressed and the placement of text and arrowheads might include a leader line.

Prompt sequence:

Command: **DIMCONT ↵**

Specify a second extension line origin or [Undo/Select] <Select>: ↵
Select continued dimension: *Select the dimension from which dimensioning will be continued*
Specify a second extension line origin or [Undo/Select] <Select>:
Dimension text = <Default>
Specify a second extension line origin or [Undo/Select] <Select>:
Dimension text = <Default>
Specify a second extension line origin or [Undo/Select] <Select>: ↵

Ordinate Dimensioning

Dimension toolbar: Ordinate

Pull-down menu: Dimension → Ordinate
Command: DIMORD ↵

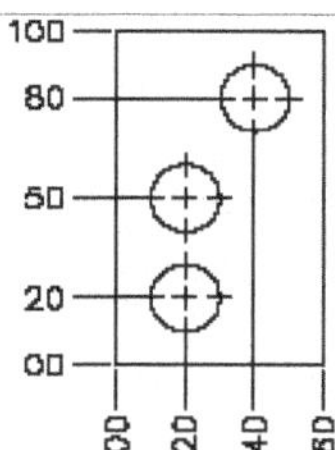

Figure 3.12 Ordinate dimensions

Ordinate dimensions measure the perpendicular distance from an origin point called the datum to a dimensioned feature, such as a hole in a part (Figure 3.12). These dimensions prevent escalating errors by maintaining accurate offsets of the features from the datum. Ordinate dimensions consist of an *X* or *Y* value with a leader line. *X*-datum ordinate dimensions measure the distance of a feature from the datum along the *X* axis. *Y*-datum ordinate dimensions measure the distance along the *Y*-axis. If you specify a point, AutoCAD automatically determines whether it is an *X*- or *Y*-datum ordinate dimension. This is called an automatic ordinate dimension. If the distance is greater for the *Y* value, the dimension measures the *X* value. Otherwise, it measures the *Y* value. AutoCAD uses the absolute coordinate value of the current UCS to determine the ordinate values. Before creating ordinate dimensions, you typically reset the UCS origin to coincide with the datum. The dimension text is aligned with the ordinate leader line, regardless of the text orientation defined by the current dimension style. You can accept the default text or supply your own.

Prompt sequence:

`Command:` **DIMORD ↵**
`Specify feature location:` *Specify the point*
`Non-associative dimension created.`
`Specify leader endpoint or [Xdatum/Ydatum/Mtext/Text/Angle]:` *Specify the point*
`Dimension text = <Default>`

Dimension Style

Style toolbar or Dimension toolbar: Dimension Style
Pull-down menu: Dimension → Style....
Command: DIMSTYLE or DDIM ↵

This command displays the Dimension Style Manager dialog box (Figure 3.13), which creates new styles, sets the current style, modifies styles, sets overrides on the current style, and compares styles. By modifying a dimension style, you can update all existing dimensions created previously with that dimension style to reflect the new settings. The Dimension Style Manager allows you to modify the following:

- Extension lines, dimension lines, arrowheads, center marks or lines, and the offsets between them.
- The positioning of the parts of the dimension in relation to one another and the orientation of the dimension text.
- The content and appearance of the dimension text.

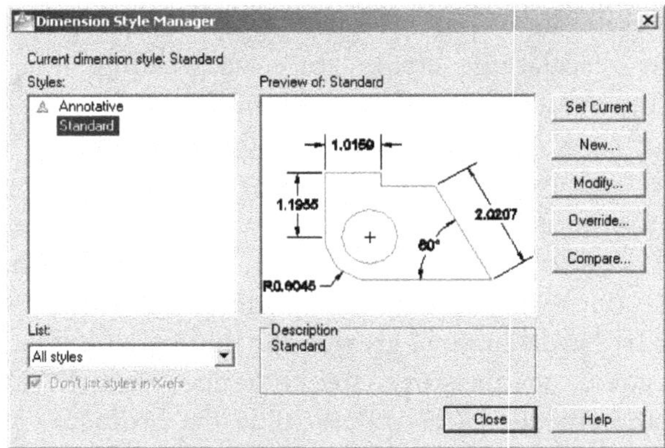

Figure 3.13 Dimension Style Manager dialog box

Current dimension style displays the current dimension style. AutoCAD assigns styles to all dimensions. If you do not change the current style, AutoCAD assigns the default STANDARD style to dimensions.

Styles displays all dimension styles in the drawing. The current style is highlighted. The item selected in list controls the dimension styles displayed. To make a style current, select it and click **Set Current**. Right-click on a style name in the Styles list to display a shortcut menu that you can use to set the current style, rename style, and delete style. You cannot delete a style that is current or in use in the current drawing.

List: Provides options of the name of the dimension styles are displayed. All Styles displays all dimension styles and Styles in Use displays only the dimension styles which are referenced by dimensions in the drawing.

Don't list styles in Xrefs: Suppresses display of dimension styles in externally referenced drawings under Styles.

Set Current: Sets the style selected under Styles to current.

New: Displays the Create New Dimension Style dialog box, in which you can define new dimension styles.

Modify: Displays the Modify Dimension Styles dialog box (Figure 3.14), in which you can modify dimension styles. Dialog box options are identical to those in the New Dimension Style dialog box.

Override: Displays the Override Current Style dialog box (Figure 3.14), in which you can set temporary overrides to dimension styles. Dialog box options are identical to those in the New Dimension Style dialog box. AutoCAD displays overrides as unsaved changes under the dimension in the Styles list.

Compare: Displays the Compare Dimension Styles dialog box, which compares the properties of two dimension styles or lists all the properties of one style.

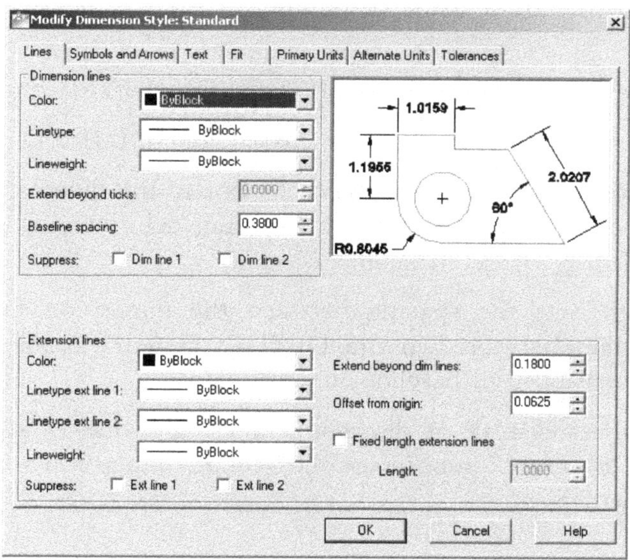

Figure 3.14 Modify Dimension Style dialog box

The New, Modify, and Override Dimension Style dialog set properties for dimension styles. After you choose Continue in the Create New Dimension Style dialog box, the New Dimension Style dialog box is displayed. You define the properties for the new style in this dialog box. The dialog box initially displays the properties of the dimension style that you selected to start the new style in the Create New Dimension Style dialog box.

Choosing either Modify or Override in the Dimension Style Manager, AutoCAD displays the Modify Dimension Style or the Override Dimension Style dialog box. The content of those dialog boxes is identical to the New Dimension Style dialog box, although you are modifying or overriding an existing dimension style rather than creating a new one.

The New, Modify, and Override Dimension Style dialog includes the following tabs: Lines and Arrows, Text, Fit, Primary Units, Alternate Units and Tolerances.

Lines tab

Lines tab sets the format and properties for dimension lines and extension lines.

Dimension lines sets the dimension line properties as follows.

Color: Sets the color for the dimension line. You can select colors from the 255 AutoCAD Color Index (ACI) colors, True Colors, and Color Book colors. If you click Select Color (at the bottom of the Color list), the Select Color dialog box is displayed. You can also enter the color name or number in the text box. (DIMCLRD system variable)

Linetype: Sets the linetype of the dimension line. (DIMLTYPE system variable)

Lineweight: Sets the lineweight of the dimension line. (DIMLWD system variable)

Extend beyond ticks: Specifies a distance to extend the dimension line passed the extension line when you use oblique, architectural, tick, integral, and no marks for arrowheads. (DIMDLE system variable)

Baseline spacing: Sets the spacing between the dimension lines of a baseline dimension. This value is stored in the DIMDLI system variable. For information about baseline dimensions, see baseline dimensioning.

Suppress: Suppresses display of dimension lines. *Dim Line 1* suppresses the first dimension line; *Dim Line 2* suppresses the second dimension line. (DIMSD1 and DIMSD2 system variables)

Extension Lines controls the appearance of the extension lines as follows.

Color: Sets the color for the extension line. You can select colors from the 255 AutoCAD Color Index (ACI) colors, True Colors, and Color Book colors. If you click Select Color (at the bottom of the Color list), the Select Color dialog box is displayed. You can also enter the color name or number in the text box. (DIMCLRE system variable)

Linetype ext 1: Sets the linetype of the first extension line. (DIMLTEX1 system variable)

Linetype ext 2: Sets the linetype of the second extension line. (DIMLTEX2 system variable)

Lineweight: Sets the lineweight of the extension line. (DIMLWE system variable)

Suppress: Suppresses the display of extension lines. Ext Line 1 suppresses the first extension line; Ext Line 2 suppresses the second extension line. (DIMSE1 and DIMSE2 system variables)

Extend beyond dim lines: Specifies a distance to extend the extension lines above the dimension line. (DIMEXE system variable)

Offset from origin: Sets the distance to offset the extension lines from the points on the drawing that define the dimension. (DIMEXO system variable)

Fixed length extension lines: Enables fixed length extension lines. (DIMFXLON system variable)

Length: Sets the total length of the extension lines starting from the dimension line toward the dimension origin. (DIMFXL system variable)

Symbols and Arrows Tab

This tab sets the format and placement for arrowheads, center marks, arc length symbols, and jogged radius dimensions.

Arrowheads controls the appearance of the dimension arrowheads as follows.

First: Sets the arrowhead for the first dimension line. When you change the first arrowhead type, the second arrowhead automatically changes to match it. (DIMBLK1 system variable)

Second: Sets the arrowhead for the second dimension line. (DIMBLK2 system variable)

Leader: Sets the arrowhead for the leader line. (DIMLDRBLK system variable)

Note: If you want to specify a user-defined arrowhead block in *First, Second* and *Leader* option, select User Arrow. The Select Custom Arrow Block dialog box is displayed. Select the name of a user-defined arrowhead block. (The block must be in the drawing.)

Arrow size: Sets the size of arrowheads. (DIMASZ system variable)

Center Marks controls the appearance of center marks and centerlines for diameter and radial dimensions. The DIMCENTER, DIMDIAMETER, and DIMRADIUS commands use center marks and centerlines. For DIMDIAMETER and DIMRADIUS, AutoCAD draws the center mark only if you place the dimension line outside the circle or arc.

Type: Provides three center mark type options: Mark, Line and None. *Mark* creates a center mark. *Line* creates a centerline. *None* creates no center mark or centerline.

Size: Sets the size of the center mark or centerline. (DIMCEN system variable)

Dimension break controls the gap width of dimension breaks.

Break size: Displays and sets the size of the gap used for dimension breaks.

Arc length symbol controls the display of the arc symbol in an arc length dimension. (DIMARCSYM system variable)

Preceding dimension text: Places arc length symbols before the dimension text. (DIMARCSYM system variable)

Above dimension text: Places arc length symbols above the dimension text. (DIMARCSYM system variable)

None: Suppresses the display of arc length symbols. (DIMARCSYM system variable)

Radius jog dimensions controls the display of jogged (zigzag) radius dimensions. Jogged radius dimensions are often created when the center point of a circle or arc is located off the page.

Jog angle: Determines the angle of the transverse segment of the dimension line in a jogged radius dimension. (DIMJOGANG system variable)

Linear jog dimensions controls the display of the jog for linear dimensions. Jog lines are often added to linear dimensions when the actual measurement is not accurately represent by the dimension. Typically the actual measurement is smaller than the desired value.

Linear jog size: Determines the height of the jog, which is determined by the distance between the two vertices of the angles that make up the jog.

Text Tab

This tab sets the format, placement, and alignment of dimension text.

Text appearance controls the dimension text format and size.

Text style: Displays and sets the current style for dimension text. Select a style from the list. To create and modify styles for dimension text, choose the [...] button next to the list. (DIMTXSTY system variable)

Text style button: Displays the Text Style dialog box, in which you can define or modify text styles.

Text color: Sets the color for the dimension text. You can select colors from the 255 AutoCAD Color Index (ACI) colors, True Colors, and Color Book colors. If you click Select Color (at the bottom of the Color list), the Select Color dialog box is displayed.

You can also enter the color name or number in the text box. (DIMCLRT system variable)

Fill color: Sets the color for the text background in dimensions. If you click Select Color (at the bottom of the Color list), the Select Color dialog box is displayed. You can also enter color name or number. (DIMTFILL and DIMTFILLCLR system variables). You can select colors from the 255 AutoCAD Color Index (ACI) colors, true colors, and Color Book colors

Text height: Sets the height of the current dimension text style. If a fixed text height is set in the Text Style (that is, the text style height is greater than 0), that height overrides the text height set here. If you want to use the height set on the Text tab, make sure the text height in the Text Style is set to 0. (DIMTXT system variable)

Fraction height scale: Sets the scale of fractions relative to dimension text. This option is available only when Fractional is selected as the Unit Format on the Primary Units tab. The value entered here is multiplied by the text height to determine the height of dimension fractions relative to dimension text. (DIMTFAC system variable)

Draw frame around text: Draws a frame around dimension text. Selecting this option changes the value stored in the DIMGAP system variable to a negative value.

Text placement controls the placement of dimension text.

Vertical: Controls the vertical placement of dimension text in relation to the dimension line. The vertical setting is stored in the DIMTAD system variable. Vertical position options include the following:

Centered: Centers the dimension text between the two parts of the dimension line.

Above: Places the dimension text above the dimension line. The distance from the dimension line to the baseline of the lowest line of text is the current text gap. See Offset from Dim Line.

Outside: Places the dimension text on the side of the dimension line farthest away from the first defining point.

JIS: Places the dimension text to conform to a Japanese Industrial Standards (JIS) representation.

Below: Places the dimension text below the dimension line.

Horizontal: Controls the horizontal placement of dimension text in relation to the dimension line and the extension lines. The horizontal setting is stored in the DIMJUST system variable. Horizontal position options include the following:

Centered: Centers the dimension text along the dimension line between the extension lines.

At ext line 1: Left-justifies the text with the first extension line along the dimension line. The distance between the extension line and the text is twice the arrowhead size plus the text gap. See Arrowheads and Offset from Dim Line.

At ext line 2: Right-justifies the text with the second extension line along the dimension line. The distance between the extension line and the text is twice the arrowhead size plus the text gap value. See Arrowheads and Offset from Dim Line.

Over ext line 1: Positions the text over or along the first extension line.

Over ext line 2: Positions the text over or along the second extension line.

Offset from dim line: Sets the current text gap, which is the distance around the dimension text when the dimension line is broken to accommodate the dimension text. AutoCAD also uses this value as the minimum length required for dimension line segments. AutoCAD positions text inside the extension lines only if the resulting segments are at least as long as the text gap. Text above or below the dimension line is placed inside only if the arrowheads, dimension text, and a margin leave enough room for the text gap. (DIMGAP system variable)

Text alignment controls the orientation (horizontal or aligned) of dimension text whether it is inside or outside the extension lines. (DIMTIH and DIMTOH system variables)

Horizontal: Places text in a horizontal position.

Aligned with dimension line: Aligns text with the dimension line.

ISO Standard: Aligns text with the dimension line when text is inside the extension lines, but aligns it horizontally when text is outside the extension lines.

Fit Tab

This tab controls the placement of dimension text, arrowheads, leader lines, and the dimension line.

Fit options controls the placement of text and arrowheads based on the space available between the extension lines. When space is available, AutoCAD places text and arrowheads between the extension lines. Otherwise, text and arrowheads are placed according to the Fit options. (DIMATFIT, DIMTIX, and DIMSOXD system variables)

Either text or arrows (best fit): Places text and arrowheads as follows: When enough space is available for text and arrowheads, places both between the extension lines. Otherwise, AutoCAD moves either the text or the arrowheads based on the best fit. When enough space is available for text only, places text between the extension lines and places arrowheads outside the extension lines. When enough space is available for arrowheads only, places them between the extension lines and places text outside the extension lines. When space is available for neither text nor arrowheads, places them both outside the extension lines.

Arrows: Places text and arrowheads as follows: When enough space is available for text and arrowheads, places both between the extension lines. When space is available for arrowheads only, places them between the extension lines and places text outside them. When not enough space is available for arrowheads, places both text and arrowheads outside the extension lines.

Text: Places text and arrowheads as follows: When space is available for text and arrowheads, places both between the extension lines. When space is available for text only, places the text between the extension lines and places arrowheads outside them. When not enough space is available for text, places both text and arrowheads outside the extension lines.

Both text and arrows: When not enough space is available for text and arrowheads, places both outside the extension lines.

Always keep text between ext lines: Always places text between extension lines. This value is stored in the DIMTIX system variable.

Suppress arrows if they don't fit inside extension lines: Suppresses arrowheads if not enough space is available inside the extension lines. (DIMSOXD system variable)

Text placement sets the placement of dimension text when it is moved from the default position, that is, the position defined by the dimension style. (DIMTMOVE system variable)

Beside the dimension line: Places dimension text beside the dimension line.

Over the dimension line, with leader: If text is moved away from the dimension line, creates a leader connecting the text to the dimension line. AutoCAD omits the leader when text is too close to the dimension line.

Over the dimension line, without leader: Keeps the dimension line in the same place when text is moved. Text that is moved away from the dimension line is not connected to the dimension line with a leader.

Scale for dimension features sets the overall dimension scale value or the paper space scaling.

Annotative: Specifies that the dimension is annotative. Click the information icon to learn more about annotative objects.

Scale dimension to layout (paper space): Determines a scale factor based on the scaling between the current model space viewport and paper space. This value is stored as 0 in the DIMSCALE system variable.

When you work in paper space, but not in a model space viewport, or when TILEMODE is set to 1, AutoCAD uses the default scale factor of 1.0 for the DIMSCALE system variable.

Use overall scale of: Sets a scale for all dimension style settings that specify size, distance, or spacing, including text and arrowhead sizes. This scale does not change dimension measurement values. This value is stored in the DIMSCALE system variable.

Fine tuning sets additional fit options.

Place text manually: Ignores any horizontal justification settings and places the text at the position you specify at the Dimension Line Location prompt. This value is stored in the DIMUPT system variable.

Draw dim line between ext lines: Draws dimension lines between the measured points even when AutoCAD places the arrowheads outside the measured points. This value is stored in the DIMTOFL system variable.

Primary Units Tab

This tab sets the format and precision of primary dimension units and sets prefixes and suffixes for dimension text.

Linear dimensions sets the format and precision for linear dimensions.

Unit format: Sets the current units format for all dimension types except Angular. This value is stored in the DIMLUNIT system variable. The relative sizes of numbers in stacked fractions are based on the DIMTFAC system variable (in the same way that tolerance values use this system variable).

Precision: Sets the number of decimal places in the dimension text. This value is stored in the DIMDEC system variable.

Fraction format: Sets the format for fractions. This value is stored in the DIMFRAC system variable.

Decimal separator: Sets the separator for decimal formats. This value is stored in the DIMDSEP system variable.

Round off: Sets rounding rules for dimension measurements for all dimension types except Angular. If you enter a value of 0.25, all distances are rounded to the nearest 0.25 unit. Similarly, if you enter a value of 1.0, AutoCAD rounds all dimension distances to the nearest integer. This value is stored in the DIMRND system variable. The number of digits displayed after the decimal point depends on the Precision setting.

Prefix: Indicates a prefix for the dimension text. You can enter text or use control codes to display special symbols (see Control Codes and Special Characters). For example, entering the control code %%c displays the diameter symbol. When you enter a prefix, it overrides any default prefixes such as those used in diameter () and radius (R) dimensioning. This value is stored in the DIMPOST system variable. If you specify tolerances, AutoCAD adds the prefix to the tolerances as well as to the main dimension.

Suffix: Indicates a suffix for the dimension text. You can enter text or use control codes to display special symbols (see Control Codes and Special Characters). For example, entering the text mm results in dimension text similar to that shown in the illustration. When you enter a suffix, it overrides any default suffixes. This value is stored in DIMPOST. If you specify tolerances, AutoCAD adds the suffix to the tolerances as well as to the main dimension.

Measurement scale: Defines measurement scale options as follows:

Scale factor sets a scale factor for linear dimension measurements. AutoCAD multiplies the dimension measurement by the value entered here. For example, if you enter 2, AutoCAD displays a one-inch dimension as two inches. The value does not apply to angular dimensions and is not applied to rounding values or to plus or minus tolerance values. This value is stored in the DIMLFAC system variable.

Apply to layout dimensions only applies the linear scale value only to dimensions created in layouts. This sets the length scale factor to reflect the zoom scale factor for objects in a model space viewport. When you select this option, the length scaling value is stored as a negative value in the DIMLFAC system variable.

Zero suppression: Controls the suppression of leading and trailing zeros, and of feet and inches that have a value of zero. Zero suppression settings also affect real-to-string conversions performed by the AutoLISP® rtos and angtos functions. AutoCAD stores this value in the DIMZIN system variable.

Leading suppresses leading zeros in all decimal dimensions. For example, 0.5000 becomes .5000.

Trailing suppresses trailing zeros in all decimal dimensions. For example, 12.5000 becomes 12.5, and 30.0000 becomes 30.

0 Feet suppresses the feet portion of a feet-and-inches dimension when the distance is less than one foot. For example, 0'-6 1/2" becomes 6 1/2".

0 Inches suppresses the inches portion of a feet-and-inches dimension when the distance is an integral number of feet. For example, 1'-0" becomes 1'.

Angular dimensions sets the current angle format for angular dimensions.

Units Format: Sets the angular units format. This value is stored in the DIMAUNIT system variable.

Precision: Sets the number of decimal places for angular dimensions. This value is stored in the DIMADEC system variable.

Zero Suppression: Suppresses leading and trailing zeros. This value is stored in DIMAZIN.

Leading suppresses leading zeros in angular decimal dimensions. For example, 0.5000 becomes .5000.

Trailing suppresses trailing zeros in angular decimal dimensions. For example, 12.5000 becomes 12.5, and 30.0000 becomes 30.

Alternate Units Tab

This tab specifies display of alternate units in dimension measurements and sets their format and precision.

Display alternate units adds alternate measurement units to dimension text. AutoCAD sets the DIMALT system variable to 1.

Alternate units sets the current alternate units format for all dimension types except Angular.

Unit format: Sets the alternate units format. This value is stored in the DIMALTU system variable. The relative sizes of numbers in stacked fractions are based on DIMTFAC (in the same way that tolerance values use this system variable).

Precision: Sets the number of decimal places in the alternate units. This value is stored in the DIMALTD system variable.

Multiplier for alternate units: Specifies a multiplier to use as the conversion factor between primary and alternate units. To determine the value of alternate units, AutoCAD multiplies all linear distances (measured by dimensions and coordinates) by the current linear scale value. The length scaling value changes the generated measurement value. The value has no effect on angular dimensions, and AutoCAD does not apply it to the rounding value or the plus or minus tolerance values. This value is stored in the DIMALTF system variable.

Round distances to: Sets rounding rules for alternate units for all dimension types except Angular. If you enter a value of 0.25, all alternate measurements are rounded to the nearest 0.25 unit. Similarly, if you enter a value of 1.0, AutoCAD rounds all dimension measurements to the nearest integer. The number of digits displayed after the decimal point depends on the Precision setting. The alternate rounding value is stored in the DIMALTRND system variable.

Prefix: Indicates a prefix for the alternate dimension text. You can enter text or use control codes to display special symbols (see Control Codes and Special Characters).

For example, entering the control code %%c displays the diameter symbol. This value is stored in the DIMAPOST system variable.

Suffix: Includes the suffix in the alternate dimension text. You can enter text or use control codes to display special symbols (see Control Codes and Special Characters). For example, entering the text cm results in dimension text similar to that shown in the illustration. When you enter a suffix, it overrides any default suffixes. This value is stored in the DIMAPOST system variable.

Zero suppression: Controls the suppression of leading and trailing zeros, and of feet and inches that have a value of zero. This value is stored in the DIMALTZ system variable.

Leading suppresses leading zeros in all decimal dimensions. For example, 0.5000 becomes .5000.

Trailing suppresses trailing zeros in all decimal dimensions. For example, 12.5000 becomes 12.5, and 30.0000 becomes 30.

0 Feet suppresses the feet portion of a feet-and-inches dimension when the distance is less than one foot. For example, 0'-6 1/2" becomes 6 1/2".

0 Inches suppresses the inches portion of a feet-and-inches dimension when the distance is an integral number of feet. For example, 1'-0" becomes 1'.

Placement controls the placement of alternate units. These values are stored in the DIMAPOST system variable.

After primary value: Places alternate units after the primary units.

Below primary value: Places alternate units below the primary units.

Tolerances Tab

This tab controls the display and format of dimension text tolerances.

Tolerance format controls the tolerance format.

Method: Sets the method for calculating the tolerance.

None: Does not add a tolerance. The DIMTOL system variable is set to 0.

Symmetrical: Adds a plus/minus expression of tolerance in which AutoCAD applies a single value of variation to the dimension measurement. A ± appears after the dimension. Enter the tolerance value in Upper Value. The DIMTOL system variable is set to 1. The DIMLIM system variable is set to 0.

Deviation: Adds a plus/minus tolerance expression. AutoCAD applies different plus and minus values of variation to the dimension measurement. A plus sign (+) precedes the tolerance value entered in Upper Value, and a minus sign (–) precedes the tolerance value entered in Lower Value. The DIMTOL system variable is set to 1. The DIMLIM system variable is set to 0.

Limits: Creates a limit dimension in which AutoCAD displays a maximum and a minimum value, one over the other. The maximum value is the dimension value plus the value entered in Upper Value. The minimum value is the dimension value minus the value entered in Lower Value. The DIMTOL system variable is set to 0. The DIMLIM system variable is set to 1.

Basic: Creates a basic dimension in which AutoCAD draws a box around the full extents of the dimension. The distance between the text and the box is stored as a negative value in the DIMGAP system variable.

Precision: Sets the number of decimal places. This value is stored in the DIMTDEC system variable.

Upper value: Sets the maximum or upper tolerance value. When you select Symmetrical in Method, AutoCAD uses this value for the tolerance. This value is stored in the DIMTP system variable.

Lower value: Sets the minimum or lower tolerance value. This value is stored in the DIMTM system variable.

Scaling for height: Sets the current height for the tolerance text. The ratio of the tolerance height to the main dimension text height is calculated and stored in the DIMTFAC system variable.

Vertical position: Controls text justification for symmetrical and deviation tolerances.

Top: Aligns the tolerance text with the top of the main dimension text. When you select this option, the DIMTOLJ system variable is set to 2.

Middle: Aligns the tolerance text with the middle of the main dimension text. When you select this option, the DIMTOLJ system variable is set to 1.

Bottom: Aligns the tolerance text with the bottom of the main dimension text. When you select this option, the DIMTOLJ system variable is set to 0.

Tolerance alignment: Controls the alignment of upper and lower tolerance values when stacked

Align decimal separators Values are stacked by their decimal separators.

Align operational symbols Values are stacked by their operational symbols.

Zero suppression: Controls the suppression of leading and trailing zeros, and of feet and inches that have a value of zero. Zero suppression settings also affect real-to-string conversions performed by the AutoLISP rtos and angtos functions. This value is stored in the DIMTZIN system variable.

Leading suppresses leading zeros in all decimal dimensions. For example, 0.5000 becomes .5000.

Trailing suppresses trailing zeros in all decimal dimensions. For example, 12.5000 becomes 12.5, and 30.0000 becomes 30.

0 Feet suppresses the feet portion of a feet-and-inches dimension when the distance is less than one foot. For example, 0'-6 1/2" becomes 6 1/2".

0 Inches suppresses the inches portion of a feet-and-inches dimension when the distance is an integral number of feet. For example, 1'-0" becomes 1'.

Alternate unit tolerance sets the precision and zero suppression rules for alternate tolerance units.

Precision: Sets the number of decimal places. This value is stored in the DIMALTTD system variable.

Zero suppression: Controls the suppression of leading and trailing zeros, and of feet and inches that have a value of zero. This value is stored in the DIMALTTZ system variable.

Leading suppresses leading zeros in all decimal dimensions. For example, 0.5000 becomes .5000.

Trailing suppresses trailing zeros in all decimal dimensions. For example, 12.5000 becomes 12.5, and 30.0000 becomes 30.

0 Feet suppresses the feet portion of a feet-and-inches dimension when the distance is less than one foot. For example, 0'-6 1/2" becomes 6 1/2".

0 Inches suppresses the inches portion of a feet-and-inches dimension when the distance is an integral number of feet. For example, 1'-0" becomes 1'.

Editing Dimensions

Dimension toolbar: Dimension Edit
Command: DIMEDIT or DIMED ↵

Once you have created a dimension, you can rotate the existing text or replace it with a new text. You can move the text to a new location or back to its home position.

```
Command: DIMEDIT ↵
Enter type of dimension editing [Home/New/Rotate/Oblique] <Home>:
```

Enter an option or press ENTER

Home option moves rotated dimension text back to its default position.

New option changes dimension text using the Multiline Text Editor. AutoCAD represents the generated measurement with angle brackets (< >). To add a prefix or a suffix to the generated measurement, enter the prefix or suffix before or after the angle brackets. To edit or replace the generated measurement, delete the angle brackets, enter the new dimension text, and then choose OK. If alternate units are not turned on in the dimension style, you can display them by entering square brackets ([ ]).

Rotate option rotates dimension text. This option is similar to the Angle option of DIMTEDIT.

Oblique option adjusts the oblique angle of the extension lines for linear dimensions. AutoCAD creates linear dimensions with extension lines perpendicular to the direction of the dimension line. The Oblique option is useful when extension lines conflict with other features of the drawing.

Editing Dimension Texts

Dimension toolbar: Dimension Text Edit
Pull-down menu: Dimension → Align Text
Command: DIMTEDIT or DIMTED ↵

The dimension text can be edited by using the DIMTEDIT command. This command is used to edit the placement and orientation of an existing associative dimension.

`Command:` **DIMTEDIT ↵**
`Select dimension:` *Select a dimension*
`Specify new location for dimension text or [Left/Right/`
`Center/Home/Angle]:` *Enter an option or press ENTER*

Location for dimension text option updates the location of the dimension text dynamically as you drag it.

Left option left-justifies the dimension text along the dimension line. This option works only with linear, radial, and diameter dimensions.

Right option right-justifies the dimension text along the dimension line. This option works only with linear, radial, and diameter dimensions.

Center option centers the dimension text on the dimension line.

Home option moves dimension text back to its default position.

Angle option changes the angle of the dimension text.

Updating Dimensions

Dimension toolbar: Dimension Update
Pull-down menu: Dimension → Update
Command: DIM; Dim: UPDATE or UP ↵

This command regenerates (updates) prevailing associative dimension objects (like arrows, text height) by using current settings for the dimension variables, dimension style, text style and units.

```
Dim: UPDATE ↵
Select objects: Select the dimension you want to update
Select objects: Select the dimension you want to update
Select objects: ↵
```

Printing a Drawing

If you want to print your drawing you have to go through PLOT command. After invoking the command Plot configuration dialog box (Figure 3.15) will appear on the screen. PLOT command specifies the device and media settings, and plots your drawing.

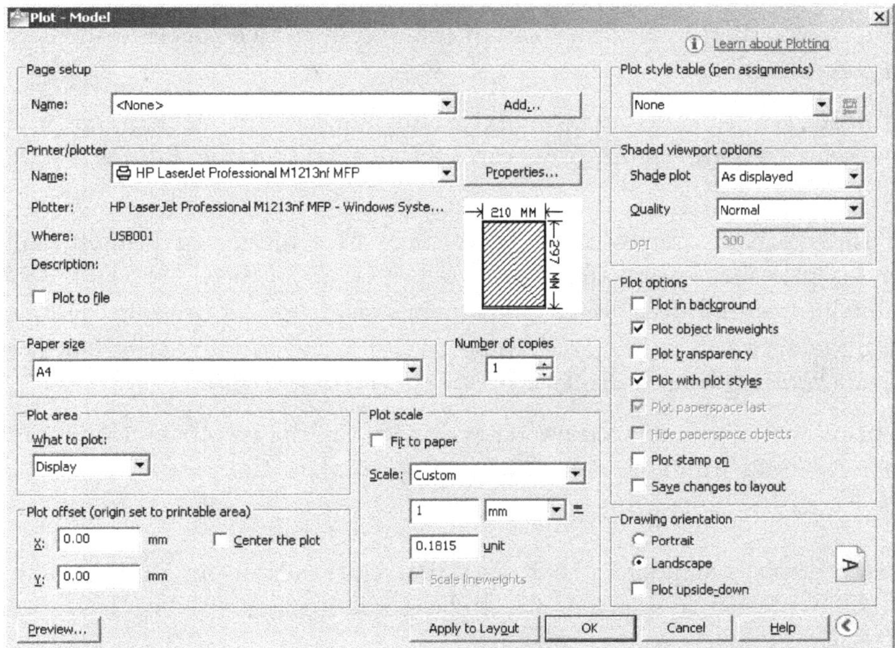

Figure 3.15 Plot configuration dialog box

Page setup displays a list of any named and saved page setups in the drawing. You can base the current page setup on a named page setup saved in the drawing, or you can create a new named page setup based on the current settings in the Plot dialog box by clicking Add.

Name: Displays the name of the current page setup.

Add: Displays the Add Page Setup dialog box, in which you can save the current settings in the Plot dialog box to a named page setup. You can modify this page setup through the Page Setup Manager.

Printer/Plotter specifies a configured plotting device to use when plotting layouts. If the selected plotter does not support the layout's selected paper size, a warning is displayed and you can select the plotter's default paper size or a custom paper size.

Name: Lists the available PC3 files or system printers from which you can select to plot the current layout. An icon in front of the device name identifies it as a PC3 file or a system printer.

Properties: Displays the Plotter Configuration Editor (PC3 editor), in which you can view or modify the current plotter configuration, ports, device, and media settings. If you make changes to the PC3 file using the Plotter Configuration Editor, the Changes to a Printer Configuration File dialog box is displayed.

Plotter: Displays the plot device specified in the currently selected page setup.

Where: Displays the physical location of the output device specified in the currently selected page setup.

Description: Displays descriptive text about the output device specified in the currently selected page setup. You can edit this text in the Plotter Configuration Editor.

Plot to file: Plots output to a file rather than to a plotter or printer. The default location for plot files is specified in the Options dialog box, Plot and Publish tab, under Default Location for Plot-to-File Operations. If the Plot to File option is turned on, when you click OK in the Plot dialog box, the Plot to File dialog box (a standard file navigation dialog box) is displayed.

Partial preview: Shows an accurate representation of the effective plot area relative to the paper size and printable area. The tooltip displays the paper size and printable area.

Paper size displays standard paper sizes that are available for the selected plotting device. If no plotter is selected, the full standard paper size list is displayed and available for selection.

Number of copies specifies the number of copies to plot. This option is not available when you plot to file.

Plot area specifies the portion of the drawing to be plotted. Under What to Plot, you can select an area of the drawing to be plotted.

Layout/Limits: When plotting a layout, plots everything within the printable area of the specified paper size, with the origin calculated from 0,0 in the layout. When plotting from the Model tab, plots the entire drawing area that is defined by the grid limits. If the current viewport does not display a plan view, this option has the same effect as the Extents option.

Extents: Plots the portion of the current space of the drawing that contains objects. All geometry in the current space is plotted. The drawing may be regenerated to recalculate the extents before plotting.

Display: Plots the view in the current viewport in the selected Model tab or the current paper space view in the layout.

View: Plots a view that was previously saved with the VIEW command. You can select a named view from the list. If there are no saved views in the drawing, this option is unavailable. When the View option is selected, a View list is displayed that lists the named views that are saved in the current drawing. You can select a view from this list to plot.

Window: Plots any portion of the drawing that you specify. When you select Window, the Window button becomes available. Click the Window button to use the pointing device to specify the two corners of the area to be plotted, or enter coordinate values.

Plot offset specifies an offset of the plotting area from the lower-left corner of the paper. In a layout, the lower-left corner of a specified plot area is positioned at the lower-left margin of the paper. You can offset the origin by entering a positive or negative value. The plotter unit values are in inches or millimeters on the paper.

Center the plot: Automatically calculates the X and Y offset values to center the plot on the paper.

X: Specifies the plot origin in the X direction.

Y: Specifies the plot origin in the Y direction.

Plot scale controls the relative size of drawing units to plotted units. The default scale setting is 1:1 when plotting a layout. The default setting is Fit to Paper when plotting from the Model tab.

Fit to paper: Scales the plot to fit within the selected paper size and displays the custom scale factor in the Scale, Inch =, and Units boxes.

Scale: Defines the exact scale for the plot. Custom defines a user-defined scale. You can create a custom scale by entering the number of inches (or millimeters) equal to the number of drawing units.

Scale lineweights: Scales lineweights in proportion to the plot scale. Lineweights normally specify the linewidth of plotted objects and are plotted with the linewidth size regardless of the plot scale.

Preview displays the drawing as it will appear when plotted on paper by executing the PREVIEW command. To exit the print preview and return to the Plot dialog box, press ESC, press ENTER, or right-click and then click Exit on the shortcut menu.

Apply to layout saves the current Plot dialog box settings to the current layout.

Plot Style Table (Pen Assignments) sets the plot style table, edits the plot style table, or creates a new plot style table.

Name (Unlabeled): Displays the plot style table that is assigned to the current Model tab or layout tab and provides a list of the currently available plot style tables.

If you select New, the Add Plot Style Table wizard is displayed, which you can use to create a new plot style table. The wizard that is displayed is determined by whether the current drawing is in color-dependent or named mode.

Edit: Displays the Plot Style Table Editor (Figure 3.16), in which you can view or modify plot styles for the currently assigned plot style table.

Shaded viewport options specifies how shaded and rendered viewports are plotted and determines their resolution level and dots per inch (dpi).

Plot options specifies options for lineweights, plot styles, shaded plots, and the order in which objects are plotted.

Drawing orientation specifies the orientation of the drawing on the paper for plotters that support landscape or portrait orientation. The paper icon represents the media orientation of the selected paper. The letter icon represents the orientation of the drawing on the page.

Portrait: Orients and plots the drawing so that the short edge of the paper represents the top of the page.

Landscape: Orients and plots the drawing so that the long edge of the paper represents the top of the page.

Plot upside-down: Orients and plots the drawing upside down.

Plot style table editor A plot style is an optional method that controls how each object or layer is plotted. Assigning a plot style to an object or a layer overrides properties such as color, lineweight, and linetype when plotting. Only the appearance of plotted objects is affected by plot style.

Plot style table lists all of the plot styles. If you select any plot style and then click on plot style table editor button, Plot Style Table Editor dialog box appears. You can use either the Table View tab or the Form View tab to adjust plot style settings. In general, the Table View tab is convenient if you have a small number of plot styles. If you have a large number of plot styles, the Form view is more convenient because the plot style names are listed at the left and the properties of the selected style are displayed to the right. You do not have to scroll horizontally to view the style and its properties. This dialog box contains the following options that may be changed:

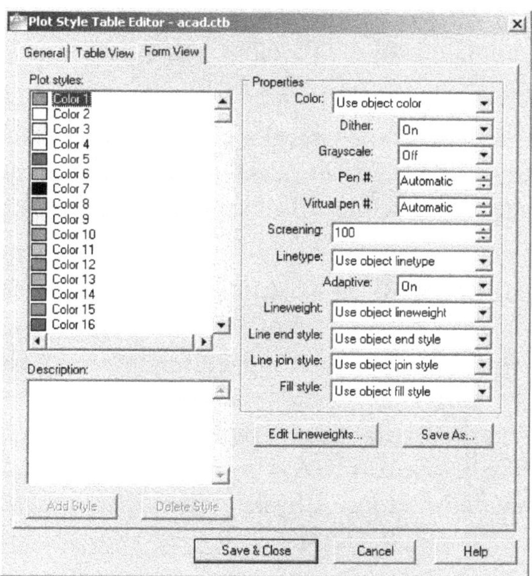

Figure 3.16 Plot Style Table Editor dialog box

Plot styles displays the names of plot styles in named plot style tables. Plot styles in named plot style tables can be changed. Plot style names in color-dependent plot style tables are tied to object color and cannot be changed. The program accepts up to 255 characters for style names. You cannot have duplicate names within the same plot style table.

Description provides a description for each plot style.

Properties specifies the settings for the new plot style you are adding to the current plot style table.

Color: Specifies the plotted color for an object. The default setting for plot style color is Use Object Color. If you assign a plot style color, the color overrides the object's color at plot time. You can choose Select Color to display the Select Color dialog box and select one of the 255 AutoCAD Color Index (ACI) colors, a true color, or a color from a color book. The color you specify is displayed in the plot style color list as Custom Color. If the plot device does not support the color you specify, it plots the nearest available color or, in the case of monochrome devices, black.

Enable dithering: Enables dithering. A plotter uses dithering to approximate colors with dot patterns, giving the impression of plotting more colors than available in the AutoCAD Color Index (ACI). If the plotter does not support dithering, the dithering setting is ignored. Dithering is usually turned off in order to avoid false line typing that results from dithering of thin vectors. Turning off dithering also makes dim colors more visible. When you turn off dithering, the program maps colors to the nearest color, resulting in a smaller range of colors when plotting. Dithering is available whether you use the object's color or assign a plot style color.

Convert to grayscale: Converts the object's colors to grayscale if the plotter supports grayscale. If you clear Convert to Grayscale, the RGB values are used for object colors. Dithering is available whether you use the object's color or assign a plot style color.

Use assigned pen number (for pen plotters only): Specifies a pen to use when plotting objects that use this plot style. Available pens range from 1 to 32. If plot style color is set to Use Object Color, or you are editing a plot style in a color-dependent plot style table, you cannot change the assigned pen number; the value is set to Automatic. If you specify 0, the field updates to read Automatic. The program determines the pen of the closest color to the object you are plotting using the information you provided under Physical Pen Characteristics in the Plotter Configuration Editor.

Virtual pen number: Specifies a virtual pen number between 1 and 255. Many non-pen plotters can simulate pen plotters using virtual pens. For many devices, you can program the pen's width, fill pattern, end style, join style, and color/screening from the front panel on the plotter. Enter 0 or Automatic to specify that the program should make the virtual pen assignment from the AutoCAD Color Index. The virtual pen setting in a plot style is used only by non-pen plotters and only if they are configured for virtual pens. If this is the case, all the other style settings are ignored and only the virtual pen is used. If a non-pen plotter is not configured for virtual pens, then the virtual and physical pen information in the plot style is ignored and all the other settings are used. You can configure your non-pen plotter for virtual pens under Vector Graphics on the Device and Document Settings tab in the PC3 Editor. Under Color Depth, select 255 Virtual Pens.

Screening: Specifies a color intensity setting that determines the amount of ink placed on the paper while plotting. The valid range is 0 through 100. Selecting 0 reduces the color to white. Selecting 100 displays the color at its full intensity. In order for screening to work, the Enable Dithering option must be selected.

Linetype: Displays a list with a sample and a description of each linetype. The default setting for plot style linetype is Use Object Linetype. If you assign a plot style linetype, the linetype overrides the object's linetype at plot time.

Adaptive adjustment: Adjusts the scale of the linetype to complete the linetype pattern. If you do not select Adaptive Adjustment, the line might end in the middle of a pattern. Turn off Adaptive Adjustment if linetype scale is important. Turn on Adaptive Adjustment if complete linetype patterns are more important than correct linetype scaling.

Lineweight: Displays a sample of the lineweight as well as its numeric value. You can specify the numeric value of each lineweight in millimeters. The default setting for plot style lineweight is Use Object Lineweight. If you assign a plot style lineweight, the lineweight overrides the object's lineweight at plot time.

Line end style: Provides the following line end styles: Butt, Square, Round, and Diamond. The default setting for Line End Style is Use Object End Style. If you assign a line end style, the line end style overrides the object's line end style at plot time.

Line join style: Provides the following line join styles: Miter, Bevel, Round, and Diamond. The default setting for Line Join Style is Use Object Join Style. If you assign a line join style, the line join style overrides the object's line join style at plot time.

Fill style: Provides the following fill styles: Solid, Checkerboard, Crosshatch, Diamonds, Horizontal Bars, Slant Left, Slant Right, Square Dots, and Vertical Bar. The default setting for Fill Style is Use Object Fill Style. If you assign a fill style, the fill style overrides the object's fill style at plot time.

Add Style button adds a new plot style to a named plot style table. The plot style is based on Normal, which uses an object's properties and does not apply any overrides by default. You must specify the overrides you want to apply after you create the new plot style. You cannot add a new plot style to a color-dependent plot style table; a color-dependent plot style table has 255 plot styles mapped to color. You also cannot add a plot style to a named plot style table that has a translation table.

Delete Style button deletes the selected style from the plot style table. Objects assigned this plot style retain the plot style assignment but plot as Normal because the plot style is no longer defined in the plot style table. You cannot delete a plot style from a named plot style table that has a translation table, or from a color-dependent plot style table.

Save As button displays the Save As dialog box and saves the plot style table to a new name.

Edit Lineweights button displays the Edit Lineweights dialog box. There are 28 lineweights available to apply to plot styles in plot style tables. If the lineweight you need does not exist in the list of lineweights stored in the plot style table, you can edit an existing lineweight. You can not add or delete lineweights from the list in the plot style table.

Edit Lineweights dialog box (Figure 3.17) modifies the values of existing lineweights.

Lineweights: Lists the lineweights in the plot style table. There are a total of 28 lineweights including Use Object Lineweight. You can modify existing lineweights, but you can not add or delete them. If you change a lineweight value, other plot styles that use the lineweight also change. When you edit a lineweight value, it is rounded and displayed with a precision of four places past the decimal point. Lineweight values must be zero or a positive number. If you create a lineweight with a zero width, the line is plotted as thin as the plotter can create it. The maximum possible lineweight value is 100 millimeters (approximately four inches).

Units for Listing: Specifies the units in which to display the list of lineweights. You can display the list of lineweights in inches or millimeters.

Edit Lineweight: Makes the selected lineweight available for editing.

Sort Lineweights: Sorts the list of lineweights by value. If you change lineweight values, choose Sort Lineweights to resort the list.

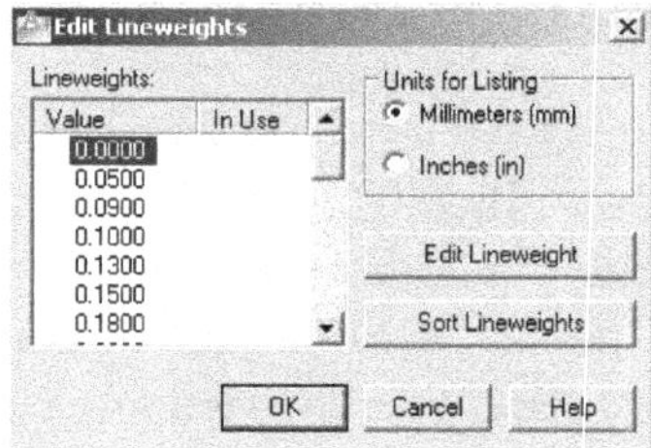

Figure 3.17 Edit Lineweights dialog box

Functional Keys

AutoCAD offers the following functional keys for various actions.

F1 key → Help
F2 key → Toggles between text window and graphic window
F3 key → Running object snap ON / OFF
F4 key → 3D object snap ON / OFF
F5 key → Toggles between different isoplanes
F6 key → Dynamic UCS ON / OFF
F7 key → Grid ON / OFF
F8 key → Ortho ON / OFF
F9 key → Snap ON / OFF
F10 key → Polar tracking ON / OFF
F11 key → Object snap tracking ON / OFF
F12 key → Dynamic input ON / OFF

Creating Command Aliases (Editing ACAD.PGP File)

A command alias is an abbreviation that you enter at the command prompt instead of entering the entire command name. For example, you can enter C instead of circle to start the CIRCLE command. An alias can be defined for any AutoCAD command, device driver command, or external command. The second section of the acad.pgp file defines command aliases.

The AutoCAD program parameters file, acad.pgp, is an ASCII text file that stores command definitions. You can change existing aliases or add new ones by editing acad.pgp in an ASCII text editor (such as Notepad). To open the PGP file:

Pull-down menu: Tools → Customize → Edit Program Parameters (acad.pgp)

Remember, before you edit acad.pgp, create a backup so that you can restore it later, if needed.

To define a command alias, add a line to the command alias section of the acad.pgp file using the following syntax:

abbreviation, **command*

where *abbreviation* is the command alias that you enter at the command prompt and *command* is the command being abbreviated. You must enter an asterisk (*) before the command name to identify the line as a command alias definition. The file can also contain comment lines preceded by a semicolon (;).

Some sample aliases for AutoCAD commands are as follows:

```
L,      *LINE
ED,     *DDEDIT
```

If you edit acad.pgp while AutoCAD is running, enter REINIT to use the revised file. You can also restart AutoCAD to automatically reload the file.

4
Advanced Techniques

Introduction

AutoCAD provides a lot of facilities for faster and smarter digital drawing. In this chapter you will learn some of those facilities which will enhance your drawing ability and performance while drawing in AutoCAD.

Creating Blocks

Draw toolbar: Make Block
Pull-down menu: Draw → Block → Make
Command: BLOCK or B ↵

Block can be defined as a group of some objects, which can be used in the drawing repeatedly. The ability to store parts of a drawing, or the entire drawing, so that they need not be redrawn when needed again in the same drawing or another drawing is extremely beneficial to the user. A block can be created by using the BLOCK command.

After invoking the BLOCK command, the Block Definition dialog box (Figure 4.1) is displayed. In the Block Definition dialog box you can enter the name of the block. After you have specified a block name, you are required to specify the insertion base point. This point is as a reference point to insert the block. After that, you are required to select the objects that will constitute the block. The Block Definition dialog box shows the following options:

Name: Names the block. The name can have up to 255 characters and can include letters, numbers, blank spaces, and any special character not used by Microsoft Windows and AutoCAD for other purposes, if the system variable EXTNAMES is set to 1. The block name and definition are saved in the current drawing.

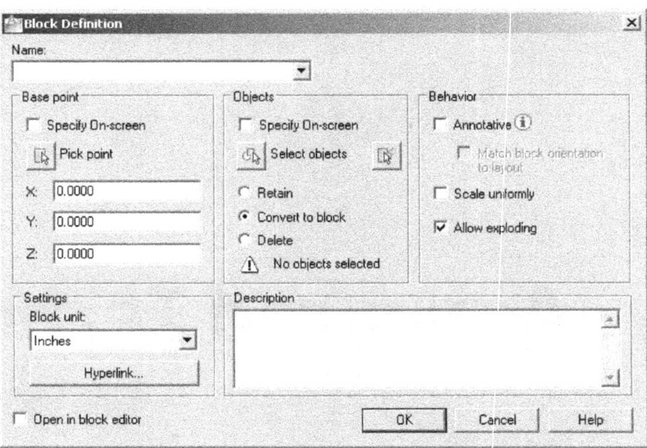

Figure 4.1 Block Definition dialog box

Preview: If an existing block is selected under Name, displays a preview of the block.

Base Point: Specifies an insertion base point for the block. The default value is 0,0,0.

X specifies the *X* coordinate value.

Y specifies the *Y* coordinate value.

Z specifies the *Z* coordinate value.

If you want to pick the point from the drawing, then click on **Pick point**. AutoCAD temporarily closes the dialog box so that you can specify an insertion base point in the current drawing.

Objects: Specifies the objects to include in the new block and whether to retain or delete the selected objects or convert them to a block instance after you create the block.

Specify on-screen prompts you to specify the objects when the dialog box is closed.

Select objects closes the Block Definition dialog box temporarily while you select the objects for the block. When you finish selecting objects, press ENTER to redisplay the Block Definition dialog box.

Quick select displays the Quick Select dialog box, which defines a selection set.

Retain option retains the selected objects as distinct objects in the drawing after you create the block.

Convert to block option converts the selected objects to a block instance in the drawing after you create the block.

Delete option deletes the selected objects from the drawing after you create the block.

Objects Selected text displays the number of selected objects.

Behavior: Specifies the behavior of the block.

Annotative specifies that the block is annotative. Click the information icon to learn more about annotative objects.

Match block orientation to layout specifies that the orientation of the block references in paper space viewports matches the orientation of the layout. This option is unavailable if the Annotative option is cleared.

Scale uniformly specifies whether or not the block reference is prevented from being non-uniformly scaled.

Allow exploding specifies whether or not the block reference can be exploded.

Settings: Specifies settings for the block.

Block unit specifies the insertion units for the block reference.

Hyperlink opens the Insert Hyperlink dialog box, which you can use to associate a hyperlink with the block definition.

Description: Specifies the text description of the block.

Open in block editor: Opens the current block definition in the Block Editor when you click OK.

Inserting Blocks

Draw toolbar: Insert Blocks
Pull-down menu: Insert → Block...
Command: INSERT or I ↵

Insertion of a predefined block is possible with the INSERT command. If you go through the INSERT command, the Insert dialog box (Figure 4.2) is displayed on the screen. Now you have to enter the name of block to be inserted, insertion point, scale, rotation etc.

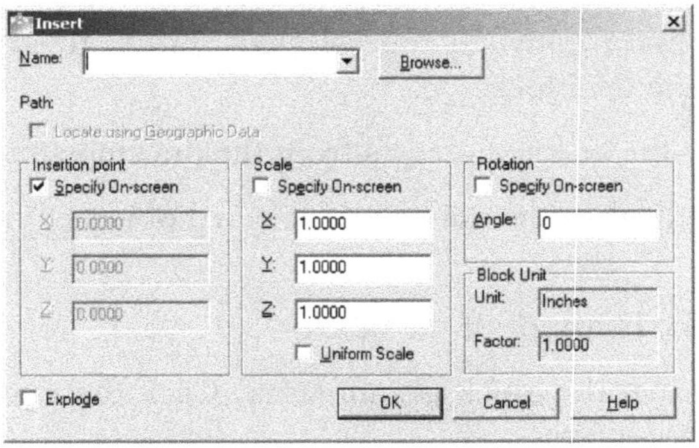

Figure 4.2 The Insert dialog box

The Insert dialog box displays the following options:

Name: Specifies the name of a block to insert, or the name of a file to insert as a block.

Browse: Opens the Select Drawing File dialog box (a standard file selection dialog box) where you can select a block or drawing file to insert.

Insertion Point: Specifies the insertion point for the block.

Specify on-screen option specifies the insertion point of the block by using the pointing device.

X sets the *X* coordinate value.
Y sets the *Y* coordinate value.
Z sets the *Z* coordinate value.

Scale: Specifies the scale for the inserted block. Specifying negative values for the X, Y, and Z scale factors inserts a mirror image of a block.

Specify on-screen option specifies the scale of the block by using the pointing device.

X sets the X scale factor.
Y sets the Y scale factor.
Z sets the Z scale factor.

Uniform Scale option specifies a single scale value for X, Y, and Z coordinates. A value specified for X is also reflected in the Y and Z values.

Rotation: Specifies the rotation angle for the inserted block in the current UCS.

Specify OnScreen option specifies the rotation of the block by using the pointing device.

Angle sets a rotation angle for the inserted block.

Block Unit: Displays information about the block units.

Unit specifies the INSUNITS value for the inserted block.

Factor displays the unit scale factor, which is calculated based on the INSUNITS value of the block and the drawing units.

Explode: Explodes the block and inserts the individual parts of the block. When Explode is selected, you can specify a uniform scale factor only.

Creating Global Block

Command: WBLOCK ↵

The blocks or symbols created by the BLOCK command can be used only in the drawing where they have created. Actually these blocks are local blocks which can be used in a single drawing file. But you may need to use a particular block in different drawings. The global blocks offer you this type of facility. Basically global block itself is a drawing file. You can create a global block by WBLOCK command. After invoking the command, the Write Block (Figure 4.3) dialog box will appear on the screen.

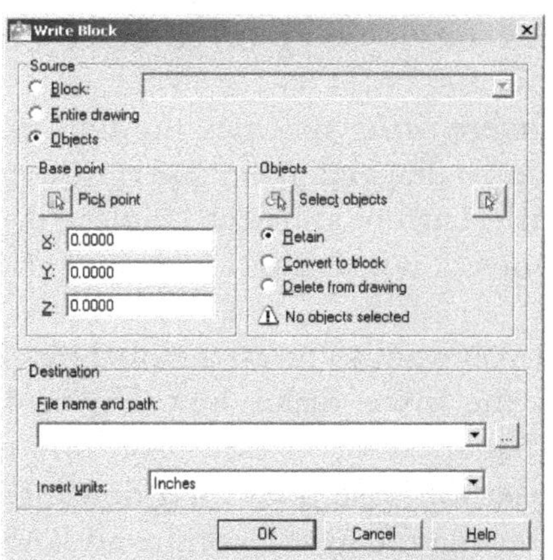

Figure 4.3 Write Block dialog box

The Write Block dialog box displays different default settings depending on whether nothing is selected, a single block is selected, or objects other than blocks are selected.

Source: Specifies blocks and objects, saves them as a file, and specifies insertion points.

Block specifies an existing block to save as a file. Select a name from the list.

Entire drawing selects current drawing as a block.

Objects specifies objects to be saved as a file.

Base Point: Specifies a base point for the block. The default value is 0,0,0.

Objects: Sets the effect of block creation on objects used to create a block.

Retain option retains the selected objects in the current drawing after saving them as a file.

Convert to block option converts the selected object or objects to a block in the current drawing after saving them as a file. The block is assigned the name in File Name.

Delete from drawing option deletes the selected objects from the current drawing after saving them as a file.

Select objects button temporarily closes the dialog box so that you can select one or more objects to save to the file.

Destination: Specifies the new name and location of the file and the units of measurement to be used when the block is inserted.
File name and path specifies a file name and path where the block or objects will be saved.
Insert units specifies the unit value to be used for automatic scaling when the new file is dragged from DesignCenter and inserted as a block in a drawing that uses different units. Select Unitless if you do not want to automatically scale the drawing when you insert it.

If you want to insert a global block in your drawing you have to click on Brows button in the Insert dialog box (Figure 4.2). All global blocks are individually a drawing file. That means you can insert any drawing file in another drawing file. By default the insertion base point of a drawing is 0,0. You can change the insertion base point of a drawing file by the command BASE.

Attribute Definition

Pull-down menu: Draw → Block → Define Attribute...
Command: ATTDEF or ATT ↵

An attribute is informational text associated with a block. An attribute definition is a template for creating an attribute; it specifies the properties of an attribute and the prompts displayed when the block is inserted. When you finish defining the attribute, the attribute tag that you specified is displayed in the drawing. When you later include the attribute tag in a block definition by using the BLOCK command, AutoCAD erases the attribute tag from the drawing if you have selected the Delete option in the Block Definition dialog box. When you insert the block, AutoCAD displays the attribute value at the same location in the block, with the same text style and alignment.

To create an attribute, you first create an attribute definition, which describes the characteristics of the attribute. The characteristics include the tag (which is a name that identifies the attribute), the prompt displayed when you insert the block, value information, text formatting, location, and any optional modes (Invisible, Constant, Verify, and Preset).

After creating the attribute definition, you select it as one of the objects when you define the block. Then, whenever you insert the block, AutoCAD prompts you with the text you specified for the attribute. For each new block insertion, you can specify a different value for the attribute. To use several attributes together, define them and then include them in the same block. For example, you can define attributes tagged 'Parts', 'Material', and 'Thickness', and then include them in a block called PARTS_DATA. If you plan to extract the attribute information for use in a parts list, you may want to keep a list of the attribute tags you have created. You will need this tag information later when you create the attribute template file.

ATTDEF command displays the **Attribute Definition** dialog box (Figure 4.4) which provides the following options:

Mode: Sets options for attribute values associated with a block when you insert the block in a drawing.

Invisible specifies that attribute values are not displayed or printed when you insert the block. ATTDISP overrides Invisible mode.

Constant gives attributes a fixed value for block insertions.

Verify prompts you to verify that the attribute value is correct when you insert the block.

Preset sets the attribute to its default value when you insert a block containing a preset attribute.

Attribute: Sets attribute data. You can enter up to 256 characters. If you need leading blanks in the prompt or the default value, start the string with a backslash (\). To make the first character a backslash, start the string with two backslashes.

Tag identifies each occurrence of an attribute in the drawing. Enter the attribute tag using any combination of characters except spaces. AutoCAD changes lowercase letters to uppercase.

Prompt specifies the prompt that is displayed when you insert a block containing this attribute definition. If you do not enter a prompt, the attribute tag is used as a prompt. If you select Constant in the Mode area, the Prompt option is not available.

Value specifies the default attribute value.

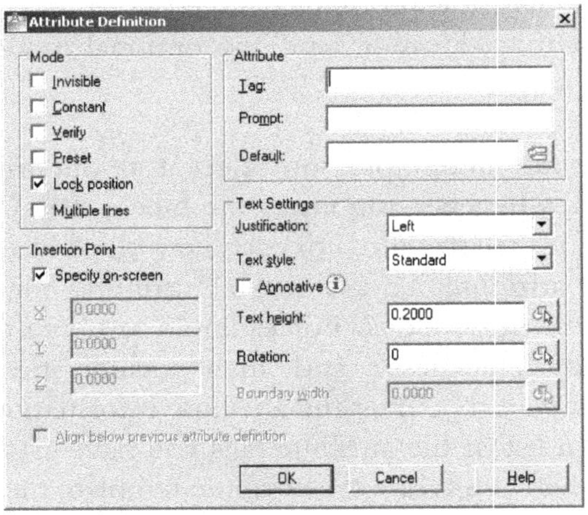

Figure 4.4 Attribute Definition dialog box

Insertion Point: Specifies the location for the attribute. Enter coordinate values or choose Pick Point and use the pointing device to specify the placement of the attribute in relation to the objects that it will be associated with.

Text Settings: Sets the justification, style, height, and rotation of the attribute text.

Justification specifies the justification of the attribute text. See TEXT for a description of the justification options.

Text Style specifies a predefined text style for the attribute text. Currently loaded text styles are displayed. To load or create a text style, see STYLE.

Annotative specifies that the attribute is annotative. If the block is annotative, the attribute will match the orientation of the block. Click the information icon to learn more about annotative objects.

Text Height specifies the height of the attribute text. Enter a value, or choose Height to specify a height with your pointing device. The height is measured from the origin to the location you specify. If you select a text style that has fixed height (anything other than 0.0), or if you select Align in the Justification list, the Height option is not available.

Rotation specifies the rotation angle of the attribute text. Enter a value, or choose Rotation to specify a rotation angle with your pointing device. The rotation angle is measured from the origin to the location you specify. If you select Align or Fit in the Justification list, the Rotation option is not available.

Boundary Width specifies the maximum length of the lines of text in a multiple-line attribute before wrapping to the next line. A value of 0.000 means that there is no restriction on the length of a line of text. Not available for single-line attributes.

Align Below Previous Attribute Definition: Places the attribute tag directly below the previously defined attribute. If you have not previously created an attribute definition, this option is not available.

Editing Attribute

Pull-down menu: Modify → Object → Attribute → Single
Command: ATTEDIT or ATE ↵

After invoking the command AutoCAD prompts to select block reference. After selecting a block the Edit Attributes dialog box is displayed. Change the values as per your requirement and click on OK.

If you enter '–ATTEDIT' at the Command prompt, ATTEDIT displays prompts in the command line; where from you can change the Value, Position, Height, Angle, Style, Layer and Color of an attribute.

`Command:` **–ATTEDIT ↵**
`Edit attributes one at a time? [Yes/No] <Y>:` *Enter y or press ENTER to edit attributes one at a time, or enter n to edit attributes globally*
`Enter block name specification <*>:` *Press ENTER, or enter a block name or a partial block name with wild-card characters (? or *) to narrow the selection to specific blocks*
`Enter attribute tag specification <*>:` *Press ENTER, or enter a tag or a partial tag with wild-card characters (? or *) to narrow the selection to specific attributes*
`Enter attribute value specification <*>:` *Press ENTER, or specify a value or a value name with wild-card characters (? or *) to narrow the selection to specific attribute values*
`Select Attributes:` *Select an attribute*
`Enter an option [Value/Position/Height/Angle/Style/Layer/Color/Next] <N>:` *Enter the property to change, or press ENTER for the next attribute*

If the original attribute was defined with aligned or fit text, the prompt does not include Angle. The Height option is omitted for aligned text. For each of the options, except Next, AutoCAD prompts for a new value. Enter the desired value and feel the changes.

Attribute Extraction

Modify II toolbar: Data Extraction
Pull-down menu: Tools → Data Extraction
Command: DATAEXTRACTION or DX ↵

You can extract attribute information from a drawing and create an external file. This feature is useful for creating parts lists with information already entered in the drawing database. Extracting attribute information does not affect the drawing.

The first time you extract data, you are prompted to save the data extraction settings in a data extraction (DXE) file. Later, if you need to edit the data extraction, you select the DXE file, which contains all the settings (data source, selected objects and their properties, output format and table style) that you used to create the extraction. For example, if you wanted to remove some property data from the extraction, you would select the DXE file that was used to create the extraction and made the desired changes. A data extraction file can also be used as a template to perform the same type of extraction in a different drawing. The DXE file stores drawing and folder selections, object and property selections, and formatting choices. If you need to extract the same type of information repeatedly, using a DXE file is time-saving and convenient. You can also edit a DXE file. You can add or remove drawings, add or remove objects, or select different properties from which to extract data. Tables that reference the same DXE file, even if those tables are in other drawings, display the changes when those tables are updated

DATAEXTRACTION command starts the Data Extraction wizard. On the Data Extraction wizard on the Begin page, click *create a new data extraction.* (If you want to use a template (DXE) file, click *use previous extraction as a template.* Click *Next.*

On the Save Data Extraction As dialog box, specify a file name for the data extraction file. Click *Save.*

On the Define Data Source page, specify the drawings or folders from which to extract data. Click *Next.*

On the Select Objects page, select the objects from which to extract data. Click *Next.*

On the Select Properties page, select the properties from which to extract data. At the right-hand side of the dialog box, category filter is there. Deselect all categories except 'Attribute' as shown in Figure 4.5. Click *Next.*

Category filter
☐ 3D Visualization
☑ Attribute
☐ Drawing
☐ General
☐ Geometry
☐ Misc

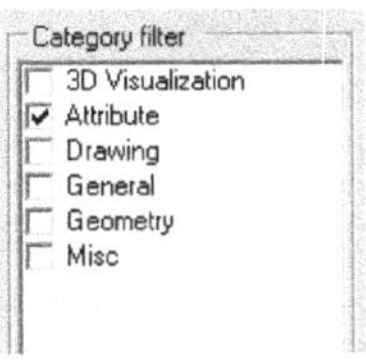

Figure 4.5 Category filter

On the Refine Data page, organize the columns if necessary. Click *Next.*

On the Choose Output page, click *output data to external file.*

On the Save As dialog specify a file name and choose a file type. Click *Save.*

On the Choose Output page, click *Next.*

On the Finish page, click Finish. Now open the saved external file using appropriate software.

Working with Design Center

Standard toolbar: DesignCenter
Pull-down menu: Tools → Palettes → DesignCenter
Command: ADCENTER or DC ↵
Shortcut key: Ctl+2

Design Centre provides you to browse, find and preview content, and inserts content into your current drawing, which includes blocks, hatches, and external references. The DesignCenter window (Figure 4.6) is divided into the tree view on the left side and the content area on the right side. Use the tree view to browse sources of content and to display content in the content area. Use the content area to add items to a drawing or to a tool palette. Below the content area, you can also display a preview or a description of a selected drawing, block, hatch pattern, or Xref. A toolbar at the top of the window provides several options and operations.

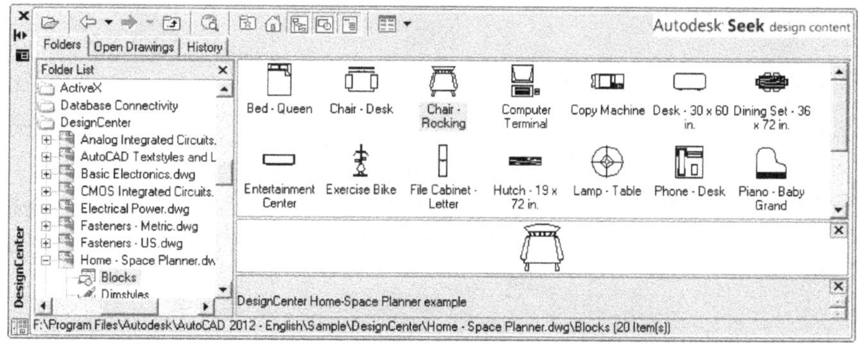

Figure 4.6 DesignCenter window

Folders Tab displays a hierarchy of navigational icons, including Networks and computers, Web addresses (URLs), Computer drives, Folders, Drawings and related support files; Xrefs, layouts, hatch styles, and named objects, including blocks, layers, linetypes, text styles, dimension styles, and plot styles within a drawing. Click an item in the tree view to display its contents in the content area. Click the plus (+) or minus (-) signs to display and hide additional levels in the hierarchy. You can also double-click an item to display deeper levels. Right-clicking in the tree view displays a shortcut menu with several related options.

The History, Open Drawings, and DC Online tabs provide alternate methods of locating content. Open Drawings displays a list of the drawings that are currently open. Click a drawing file and then click one of the definition tables from the list to load the content into the content area. History displays a list of files opened previously with DesignCenter. Double-click a drawing file from the list to navigate to the drawing file in the tree view of the Folders tab and to load the content into the content area. DC Online provides content from the DesignCenter Online web page including blocks, symbol libraries, manufacturer's content, and online catalogs.

Bookmark Frequently-Used Content

DesignCenter provides a solution to finding content that you need to access quickly on a regular basis. Both the tree view and the content area include options that activate a folder called Favorites. The Favorites folder can contain shortcuts to content on local or network drives as well as in Internet locations. When you select a drawing, folder, or another type of content and choose Add to Favorites, a shortcut to that item is added to the Favorites folder. The original file or folder does not actually move; in fact, all the shortcuts you create are stored in the Favorites folder. The shortcuts saved in the Favorites folder can be moved, copied, or deleted by using Windows Explorer.

Working with Tool Palettes

Standard toolbar: Tool Palettes Window
Pull-down menu: Tools → Palettes → Tool Palettes
Command: TOOLPALETTES or TP ↵
Shortcut key: Ctl+3

Tool palettes are tabbed areas within the Tool Palettes window (Figure 4.7) that provide an efficient method for organizing, sharing, and placing blocks and hatches. Drag blocks and hatches from a tool palette to place those objects quickly on a drawing. Tool palettes can also contain custom tools provided by third-party developers.

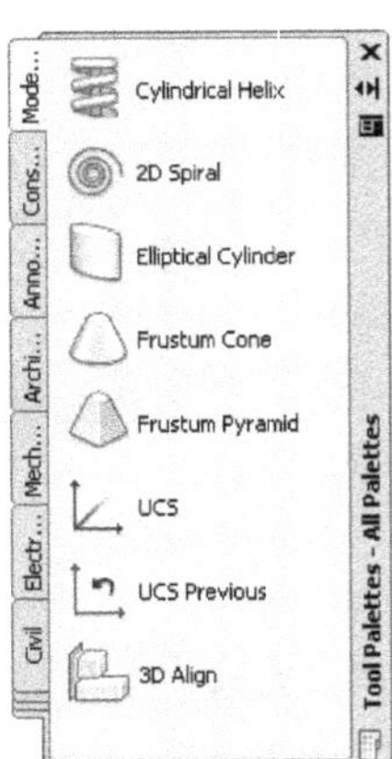

Figure 4.7 Tool Palettes window

Add tools to a tool palette with the following methods:

If *DesignCenter* is not already open, press Ctl+2. In the *DesignCenter* tree view or the content area, right-click a folder or drawing file or block. On the shortcut menu, click **Create Tool Palette**. A new tool palette is created that contains all the blocks and hatches in the selected folder or drawing.

Once tools are placed in a tool palette, you can rearrange them by dragging them within the tool palette. A tool palette tab can be moved up and down the list of tabs from the tool palette shortcut menu, or from the Tool Palettes tab of the Customize dialog box. Similarly, you can delete tool palettes that are needed no longer. Deleted tool palettes are lost unless they are first saved by exporting them to a file. You can control the path to your tool palettes on the Files tab in the Options dialog box. This path can be to a shared network location.

External References

Reference toolbar: External References
Pull-down menu: Insert → External References...
Command: XREF or XR ↵

AutoCAD treats an external reference (commonly called Xref) as a type of block definition with some important differences. When you insert a drawing as a block reference, it is stored in the drawing and is not updated if the original drawing changes. When you attach a drawing as an Xref, you link that referenced drawing to the current drawing; any changes to the referenced drawing are displayed in the current drawing when it is opened. A drawing can be attached as an external reference to multiple drawings at the same time. Conversely, multiple drawings can be attached as external references to a single drawing. The saved path used to locate the external reference can be an absolute (fully specified) path, a relative (partially specified) path, or no path. If an external reference contains any variable block attributes, AutoCAD ignores them. Remember, external references must be model space objects. They can be attached at any scale, location, and rotation.

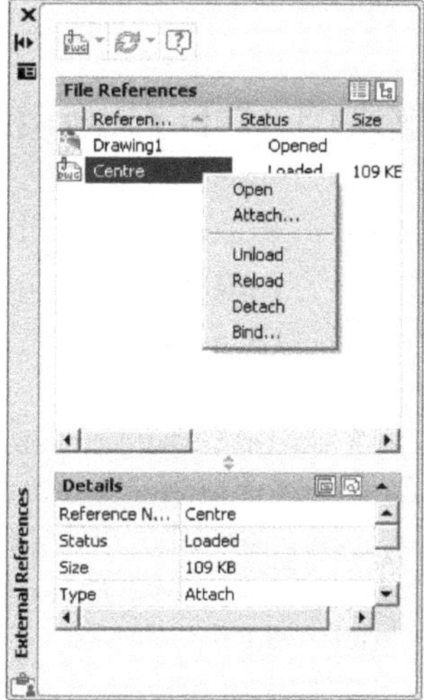

Figure 4.8 External References palette

After invoking the command the *External Reference* palette (Figure 4.8) displays on the screen. The External References palette organizes, displays, and manages referenced files, such as referenced drawings (Xrefs), attached DWF, DWFx, or DGN underlays, and imported raster images. Only DWG, DWF, DWFx, and raster image files can be opened directly from the External References palette.

The External References palette contains several buttons, and is split into two panes. The upper pane, called the File References Pane, can display file references in a list or in a tree structure. Shortcut menus and function keys provide options for working with the files. The lower pane, called the Details/Preview Pane, can display properties for the selected file references or it can display a thumbnail preview of the selected file reference.

Attach (file type) button: The Attach button displays a list of file types that you can attach. The following options are displayed:

Attach DWG Starts the XATTACH command.

Attach Image Starts the IMAGEATTACH command.

Attach DWF Starts the DWFATTACH command.

Attach DGN Starts the DGNATTACH command

Refresh/Reload all references button: The following options are available:

Refresh Resynchronizes the status data of referenced drawing files with the data in memory. Refresh interacts primarily with Autodesk Vault.

Reload all references Updates all file references to ensure that the most current version is used. Updating also occurs when you first open a drawing that contains file references.

XATTACH Command

Reference toolbar: Attach Xref
Pull-down menu: Insert → External References...
Command: XATTACH or XA ↵

This command attaches a drawing as an external reference (xref). If you attach a drawing that itself contains an attached xref, the attached xref appears in the current drawing. Similar to blocks, attached xrefs can be nested. If another person is currently editing the xref, the drawing attached is based on the most recently saved version. Once you invoke XATTACH command, the *Select Reference* dialog box (a standard file selection dialog box) appears on the screen, in which you can select xref files for the current drawing. After selecting the xref file and clicking on Open in the Select Reference dialog box, the *Attach External Reference* dialog box (Figure 4.9) appears on the screen. This dialog has the following options.

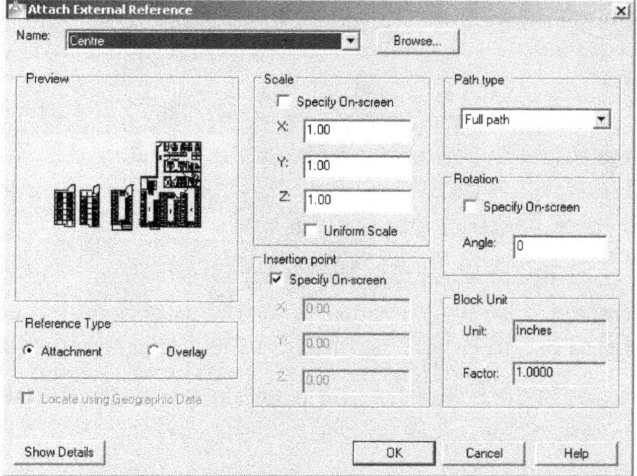

Figure 4.9 Attach External References dialog box

Name: Displays the xref name in the list after an xref is attached. When an attached xref is selected in the list, its path is displayed in Path.

Browse: Choose Browse to display the Select Reference dialog box, in which you can select new xrefs for the current drawing.

Found In: Displays the path where the xref was found.
If no path was saved for the xref or if the xref is no longer located at the specified path, the program searches for the xref in the following order:

Current folder of the host drawing
Project search paths defined on the Files tab in the Options dialog box and in the PROJECTNAME system variable
Support search paths defined on the Files tab in the Options dialog box
Start-in folder specified in the Microsoft® Windows® application shortcut

Saved Path: Displays the saved path, if any, that is used to locate the xref. This path can be a full (absolute) path, a relative (partially specified) path, or no path.

Reference Type: Specifies whether the xref is an attachment or an overlay. Unlike an xref that is an attachment, an overlay is ignored when the drawing to which it is attached is then attached as an xref to another drawing.

Path Type: Specifies whether the saved path to the xref is the full path, a relative path, or no path. You must save the current drawing before you can set the path type to Relative Path. For a nested xref, a relative path always references the location of its immediate host and not necessarily the currently open drawing. The Relative Path option is not available if the referenced drawing is located on a different local disk drive or on a network server.

Insertion Point: Specifies the insertion point of the selected xref.
Specify On-Screen displays command prompts and makes the X, Y, and Z options unavailable. The descriptions of the options displayed at the command prompt match the descriptions of the corresponding options under Insertion Point in INSERT.

X specifies the *X*-coordinate value for the point at which an xref instance is inserted into the current drawing. The insertion point in the current drawing is aligned with the insertion point defined in the BASE command in the referenced file.

Y specifies the *Y*-coordinate value for the point at which an xref instance is inserted into the current drawing. The insertion point in the current drawing is aligned with the insertion point defined in the BASE system variable in the referenced file.

Z specifies the Z-coordinate value for the point at which an xref instance is inserted into the current drawing. The insertion point in the current drawing is aligned with the insertion point defined in the BASE system variable in the referenced file.

Scale: Specifies the scale factors for the selected xref.

Specify On-Screen displays command prompts and makes the X, Y, and Z Scale Factor options unavailable.

X Scale Factor specifies the X scale factor for the xref instance.

Y Scale Factor specifies the Y scale factor for the xref instance.

Z Scale Factor specifies the Z scale factor for the xref instance.

Uniform Scale ensures that the Y and Z scale factors are equal to the X scale factor.

Rotation: Specifies the rotation angle for the xref instance.

Specify On-Screen displays command prompts and makes the Angle option unavailable. You are prompted for a rotation angle, as described in INSERT.

Angle specifies the rotation angle at which an xref instance is inserted into the current drawing.

Block Unit: Displays information about the block units.

Unit displays the specified INSUNITS value for the inserted block.

Factor displays the unit scale factor, which is calculated based on the INSUNITS value of the block and the drawing units.

INSUNITS command specifies a drawing-units value for automatic scaling of blocks, images, or xrefs inserted or attached to a drawing (for example, 0 for Unspecified (No units), 1 for Inches, 2 for Feet, 3 for Miles, 4 for Millimeters, 5 for Centimeters, 6 for Meters, 7 for Kilometers, 8 for Microinches, etc.)

Reference toolbar: Attach Xref
Pull-down menu: Insert → External References...
Command: XATTACH or XA ↵

IMAGE Command

Reference toolbar: Attach Image
Pull-down menu: Insert → Raster Image References...
Command: IMAGEATTACH or IAT ↵

Raster images consist of a rectangular grid of small squares or dots known as pixels. For example, a photograph of a house is made up of a series of pixels colorized to represent the appearance of a house. A raster image references the pixels in a specific grid. Designers and manufactures store images of their designs or products. Raster can be generated by scanning a paper or photo. Raster files take a large amount of memory space in the computer and it is very difficult to edit a raster image. To solve this problem you can convert the raster image to editable vector file. This technology is known as Raster-to-vector (R2V) conversion or in simple vectorization. Nowadays, this type of conversion is very popular in CAD field. The procedure includes: attachment of image, scaling it properly and vectorization (redrawing the entities following the raster). In computer graphics, vectorization refers to the process of using software and hardware technology/services to convert raster graphics into vector graphics

After invoking the command, the *Select Image File* dialog box (a standard file selection dialog box) is displayed. Once you select an image file, the *Image* dialog box is displayed. This dialog box is similar to the *Attach External Reference* dialog box (Figure 4.9) and all the options and operations are also similar. Once you attach the image, you can use line/pline/circle/arc or any other drawing commands to draw the vector graphics over the raster image.

DRAWORDER Command

Draw Order toolbar:
Pull-down menu: Tools → Draw Order → Bring to Front / Sent to Back / ....
Command: DRAWORDER or DR ↵

DRAWORDER changes the drawing and plotting order of any object in the AutoCAD drawing database. In addition to moving objects to the 'front' or 'back' of the sort order, you can order objects relative to another object; i.e., above or below a selected object.

```
Command: DRAWORDER ↵
Select objects: Use an object selection method
Enter object ordering option [Above object/Under object/
Front/Back] <Back>: Enter an option and press ENTER
```

Above object option: Moves the selected object above a specified reference object.

Under object option: Moves the selected object below a specified reference object.

Front option: Moves the selected object to the top of the order of objects in the drawing.

Back option: Moves the selected object to the bottom of the order of objects in the drawing.

Concept of Slide

A slide is a snapshot of a drawing. Although it contains a picture of the drawing at a given instant, it is not a drawing file. You can neither import a slide file into the current drawing, nor can you edit or print a slide. You can only view it. You can use slide files for making presentations within AutoCAD and for viewing a snapshot of a drawing while working on a different drawing.

You can create a slide by saving the current view in slide format (MSLIDE command). A slide created in model space shows only the current viewport. A slide created in paper space shows all visible viewports and their contents. Slides show only what was visible. They do not show objects on layers that were turned off or frozen or objects in viewports which were turned off. When you view a slide file, it temporarily replaces objects on the screen (VSLIDE command). You can draw on top of it, but when you change the view (by redrawing, panning, or zooming), the slide file disappears, and AutoCAD redisplays only what you drew and any preexisting objects. You can display slides one by one or use a script to display slides in sequence. You cannot edit a slide. You must change the original drawing and remake the slide.

MSLIDE Command

Command: MSLIDE ↵

After entering the command the *Create Slide File* dialog box (a standard file selection dialog box) is displayed. Enter a file name or select a slide (SLD) file from the list. A slide file is a raster image of a viewport.

VSLIDE Command

Command: VSLIDE ↵

The Select Slide File dialog box, a standard file selection dialog box, is displayed once you enter VSLIDE command. Enter a slide file name (.sld extension) to display. When you press ENTER or choose Open, the slide file is opened in AutoCAD.

The REDRAW (or R) command vanishes the slide from the screen and returns the present drawing space.

Script File

A script is a text file with one command on each line. You can invoke a script at startup, or you can run a script from within AutoCAD by using the SCRIPT command. A script provides an easy way to create continuously running displays for product demonstrations and trade shows. You can create script files outside of AutoCAD using a text editor (such as Notepad) or a word processor (such as Microsoft Word) that can save the file in ASCII format. The file extension must be **.SCR**.

Each line of the script file contains a command. Each blank space in a script file is significant. AutoCAD accepts either a space or ENTER as a command or data field terminator. You must be very familiar with the sequence of AutoCAD prompts to provide an appropriate sequence of responses in the script file.

A script can execute any command at the Command prompt except a command that displays a dialog box. AutoCAD also provides command line versions of the dialog box commands.

Script files can contain comments. Any line that begins with a semicolon (;) is considered a comment, and AutoCAD ignores it while processing the script file. The last line of the file must be blank.

The following commands are useful in scripts:

DELAY - Provides a timed pause within a script (in milliseconds)

GRAPHSCR - Switches from the text window to the drawing area

RESUME - Continues an interrupted script

RSCRIPT - Repeats a script file

TEXTSCR - Switches to the text window

Remember, AutoCAD considers a script to be a group, a unit of commands, reversible by a single U command.

SCRIPT Command

Pull-down menu: Tools → Run Script
Command: SCRIPT or SCR ↵

AutoCAD displays the Select Script File dialog box (a standard file selection dialog box). Enter the file name of a script to run it.

PURGE Command

Pull-down menu: File → Drawing Utilities → Purge
Command: PURGE or PU ↵

To remove unused named objects, including block definitions, dimension styles, layers, linetypes, and text styles, use PURGE command. After entering the command the *Purge* dialog box is displayed. Select any item (Block or Layer or Linetype etc.) and click on *Purge* or for all items click on *Purge All.*

Template Drawing

A drawing template file contains standard settings. Select one of the template files supplied, or create your own template files. Drawing template files have a .dwt file extension. When you create a new drawing based on an existing template file and make changes, the changes in the new drawing do not affect the template file. You can use one of the template files supplied with AutoCAD, or you can create your own template files.

When you need to create several drawings that use the same conventions and default settings, you can save time by creating or customizing a template file instead of specifying the conventions and default settings each time you start. Conventions and settings are commonly stored in template files:

Unit type and precision
Title blocks, borders, and logos
Layer names
Snap, Grid, and Ortho settings
Drawing (grid) limits
Dimension styles
Text styles
Linetypes

By default, drawing template files are stored in the template folder, where from they are easily accessible. However, you can save the template in your desired folder as well.

> **Note:** You can start a new drawing with the original defaults by using NEW to display the Select Template dialog box. To do this, click the arrow next to the Open button and then click one of the 'no template' options from the list.

CAL Command

Command: CAL (or 'CAL for transparent use) ↵

CAL is an online geometry calculator that evaluates point (vector), real, or integer expressions. The expressions can access existing geometry using the object snap functions such as CEN, END, and INS. By entering a formula on the command line, you can quickly solve a mathematical problem or locate points in your drawing. CAL evaluates expressions according to standard mathematical rules of precedence. The CAL command runs the AutoCAD 3D calculator utility to evaluate vector expressions (combining points, vectors, and numbers) and real and integer expressions. The calculator performs standard mathematical functions. It also contains a set of specialized functions for calculations involving points, vectors, and AutoCAD geometry. With the CAL command, you can (1) Calculate a vector from two points, the length of a vector, a normal vector (perpendicular to the *XY*-plane), or a point on a line; (2) Calculate a distance, radius, or angle; (3) Specify a point with the pointing device; (4) Specify the last-specified point or intersection; (5) Use object snaps as variables in an expression; (6) Convert points between a UCS and the WCS; (7) Filter the *X*, *Y*, and *Z* components of a vector; (8) Rotate a point around an axis.

You can use CAL whenever you need to calculate a point or a number within an AutoCAD command. For example, you enter (mid+cen)/2 to specify a point halfway between the midpoint of a line and the center of a circle.

QUICKCALC Command

Standard toolbar: QuickCalc
Pull-down menu: Tools → Palettes → QuickCalc
Command: QUICKCALC or QC ↵
Shortcut menu: Right-click and click QuickCalc

The QuickCalc calculator (Figure 4.10) includes basic features similar to most standard mathematical calculators. In addition, QuickCalc has features specific to AutoCAD such as geometric functions, a Units Conversion area, and a Variables area.

Unlike most calculators, QuickCalc is an expression builder. For greater flexibility, it does not immediately calculate an answer when you click a function. Instead, you compose an expression that you can easily edit and, when you are finished, you click the equal sign (=) or press ENTER. Later, you can retrieve the expression from the History area, modify it, and recalculate the results.

Once you enter the QuickCalc command the QuickCalc calculator palette (Figure 4.10) displays on the screen. Click the More/Less button on the calculator. You can use the expand/collapse arrows to open and close areas. You can also control the size, location, and appearance of QuickCalc. Now use your keyboard and/or mouse to perform calculations.

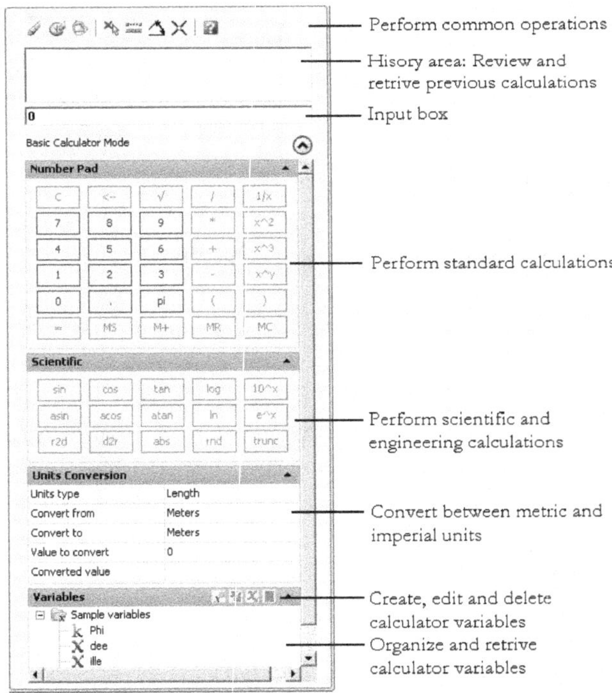

Figure 4.10 QuickCalc calculator palette

5

Isometric and 3D Drawings

Introduction

AutoCAD is an excellent tool for isometric and three dimensional (3D) drawings. AutoCAD supports three types of 3D drawings: wireframe, surface, and solid. Each type has its own creation and editing techniques. You can display parallel and perspective views with several commands to facilitate constructing and visualizing 3D models.

A wireframe model is a skeletal description of a 3D object. Wireframe models are not solid objects. Isometric drawing is basically 3D views of 2D objects. There are no surfaces in a wireframe model; it consists only of points, lines, and curves that describe the edges of the object. With AutoCAD you can create wireframe models by positioning 2D (planar) objects anywhere in 3D space. AutoCAD also provides some 3D wireframe objects, such as 3D polylines (that can only have a CONTINUOUS linetype) and splines. Because each object that makes up a wireframe model must be independently drawn and positioned, this type of modelling can be the most time-consuming.

Surface modelling is more sophisticated than wireframe modelling in that it defines not only the edges of a 3D object, but also its surfaces. The AutoCAD surface modeller defines faceted surfaces by using a polygonal mesh. Because the faces of the mesh are planar, the mesh can only approximate curved surfaces.

Solid modelling is the easiest type of 3D modelling to use. With the AutoCAD solid modeler, you can make 3D objects by creating basic 3D shapes: boxes, cones, cylinders, spheres, wedges, and tori (donuts). You can then combine these shapes to create more complex solids by joining or subtracting them or finding their intersecting (overlapping) volume. You can also create solids by sweeping a 2D object along a path or revolving it about an axis. Remember, solid models have several advantages over wireframes or surface models.

Before you start 3D modelling you are required to know some basic concepts of isometric drawing, and how to draw it by using AutoCAD.

Isometric Drawing

Used to help in visualizing shape of an object.
Can be read by individual who lacks knowledge of orthographic representation.
Isometric means equal measures.
Three angles between three principal axes of an isometric drawing are equal, i.e. 120°.

Isometric view of an object is obtained by:
Rotation object by 450° about vertical axis.
Tilting object forward about horizontal axis through 350° (precisely 350° 16').
Thus projection obtained in vertical plane will be isometric projection.

Lengths of the sides are shortened to 82% of their true length. By convention, isometric drawings are always drawn to full scale.

In the AutoCAD environment
Isometric drawing is a 2D drawing
They are drawn in 2D plane
These drawings should not be confused with a 3D drawing.

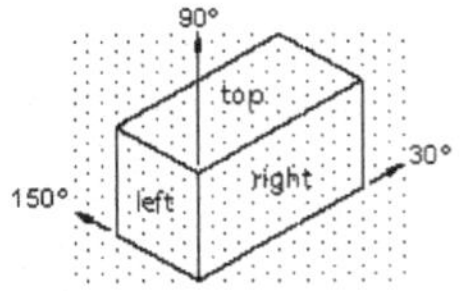

Isometric drawings have 3 axes
1. Right horizontal axis
2. Vertical axis
3. Left horizontal axis

Isometric axis and planes are shown in Figure 5.1.

90°
top
right
30°
150°
left

Figure 5.1 Three principal isometric drawing planes

Angles do not appear true in isometric drawing.

Setting Isometric Grid and Snap

```
Command: SNAP ↵
Specify snap spacing or [ON/OFF/Aspect/Rotate/Style/
Type] <0.5000>: S ↵
Enter snap grid style [Standard/Isometric] <S>: I ↵
Specify vertical spacing <0.5000>: Enter your desired value
```

Your cross-hair courser will look like the picture shown in Figure 5.2. Grid lines may not be displayed initially. Turn the grids on using GRID command or press function key F7.

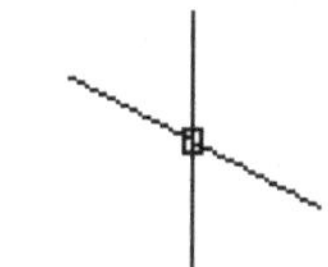

Figure 5.2 Isometric style of cross-hair cursor

You can not set the aspect ratio for isometric grid. The grid is a rectangular pattern of lines. Using the grid is similar to placing a sheet of grid paper under a drawing. The grid helps you align objects and visualize the distances between them. The grid is not plotted. If you zoom in or out of your drawing, you may need to adjust grid spacing to be more appropriate for the new magnification. The cross-hair courser orientation depends on current *isoplane* setting. You can toggle among isoplane right, left or top by pressing F5 key or ISOPLANE command.

Isometric Circle Using ELLIPSE Command

```
Command: ELLIPSE ↵
Specify axis endpoint of ellipse or [Arc/Center/Isocircle]: I ↵
Specify center of isocircle: Specify the center
Specify radius of isocircle or [Diameter]: Enter the radius or D for diameter
```

Remember, you cannot draw isometric circle if isometric snap is not on.

Dimensioning Isometric Objects

At first dimension the drawing by using standard dimensioning command then edit the dimensions to change the dimensions to oblique dimensions. The following example makes the dimension 30° oblique. You can use DIMEDIT command also (refer Chapter 3).

```
Command: DIM ↵
Dim: OBLIQUE or OB ↵
Select objects: Select the dimension
Select objects: ↵
Enter obliquing angle (press ENTER for none): 30 ↵
```

Wireframe Modelling

Wireframe models can be created by using simple AutoCAD commands like LINE, ARC, CIRCLE, PLINE etc. You can use 3D coordinates to generate the command or you can set the elevation (the bottom location of the object from XY plane) and thickness (the actual height or thickness of the object) if you want to use 2D coordinates, or you can use both.

Command: **ELEV** ↵
Specify new default elevation <0.0000>: *Specify the elevation*
Specify new default thickness <0.0000>: *Specify the height*

Using 3D coordinates

Using absolute coordinates

Command: **LINE** ↵
Specify first point: 2,0,0 ↵ (*X,Y,Z coordinates*)
Specify next point or [Undo]: 8,0,0 ↵
Specify next point or [Undo]: ↵

Using relative coordinates

Command: **LINE** ↵
Specify first point: **2,0,0** ↵
Specify next point or [Undo]: **@6,0,0** ↵
Specify next point or [Undo]: ↵

VPOINT Command

Pull-down menu: View → 3D Views → Viewpoint
Command: VPOINT ↵

You can define a viewing direction by entering the coordinate values of a point or the measures of two angles of rotation. This point represents your position in 3D space as you view the model while looking toward the origin (0,0,0).

Command: **VPOINT** ↵
Current view direction: VIEWDIR=0.0000,0.0000,1.0000
Specify a view point or [Rotate] <display compass and tripod>: **-1,-1,1** ↵

Surface Modelling

AutoCAD offers a wide range of commands to create surfaces. These surfaces are basically wire mesh. A mesh represents an object's surface by using planar facets. You can create meshes in both 2D and 3D, but they are used primarily for 3D.

RULESURF Command

With the command RULESURF, you can create a surface mesh between two objects (Figure 5.3). You use two different objects to define the edges of the ruled surface: lines, points, arcs, circles, ellipses, elliptical arcs, 2D polylines, 3D polylines, or splines.

Command: **RULESURF ↵**
Current wire frame density: SURFTAB1=6
Select first defining curve: *Select first object*
Select second defining curve: *Select second object*

You can specify any two points on closed curves to complete RULESURF. For open curves, AutoCAD starts construction of the ruled surface based on the locations of the specified points on the curves (Figure 5.4).

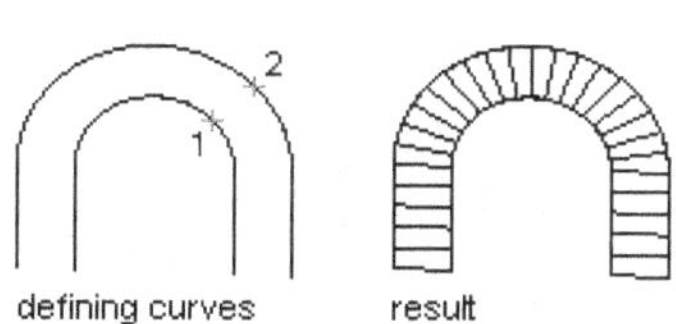

Figure 5.3 Ruled surface

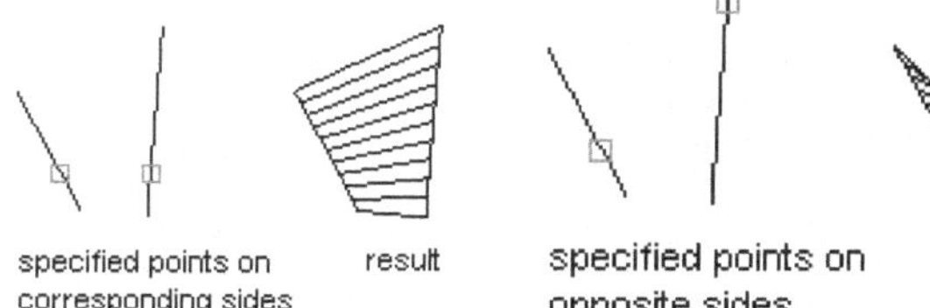

Figure 5.4 Selection procedure for ruled surface

Number of lines in mesh is controlled by system variable SURFTAB. Value of SURFTAB1 can vary from 3 to 1024. Default value is 6. You can change the value by the command SURFTAB1.

Command: **SURFTAB1 ↵**
Enter new value for SURFTAB1 <6>: *Specify new value*

TABSURF Command

With the TABSURF command, you can create a surface mesh representing a general tabulated surface defined by a path curve and a direction vector. The path curve can be a line, arc, circle, ellipse, elliptical arc, 2D polyline, 3D polyline, or spline. The direction vector can be a line or an open 2D or 3D polyline. TABSURF creates the mesh as a series of parallel polygons running along a specified path. You must have the original object and the direction vector which has already been drawn (Figure 5.5).

Command: **TABSURF ↵**
Current wire frame density: SURFTAB1=6
Select object for path curve: *Select the curve which will generate the surface*
Select object for direction vector: *Select the direction vector*

End point closest to your pick point on the direction vector will be the base point of extrusion and the opposite end of the direction vector indicates the direction of the extrusion. You can erase the direction vector once you have created the mesh. The SURFTAB1 command, as discussed earlier, controls the number of lines in a mesh.

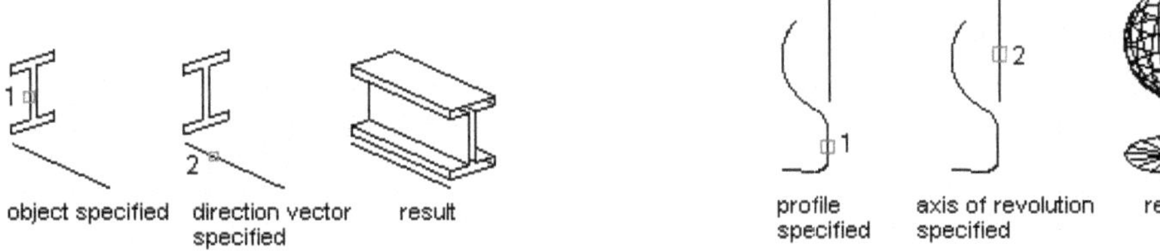

Figure 5.5 Tabulated surface

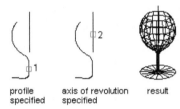

Figure 5.6 Revolved surface

REVSURF Command

Use the REVSURF command to create a surface of revolution by rotating a profile of the object about an axis (Figure 5.6). REVSURF is useful for surfaces with rotational symmetry.

Command: **REVSURF ↵**
Current wire frame density: SURFTAB1=6 SURFTAB2=6
Select object to revolve: *Select the object which will generate the surface*
Select object that defines the axis of revolution: *Select the axis*
Specify start angle <0>: *Enter an angle or press ENTER*
Specify included angle (+=ccw, -=cw) <360>: *Enter an angle or press ENTER*

SURFTAB1 and SURFTAB2 system variables control the mesh density. Do not select a too high value for these variables; otherwise the file size will be increased dramatically.

EDGESURF Command

With the EDGESURF command, you can create a coons surface patch mesh, as shown in Figure 5.7, from four objects called edges. Edges can be arcs, lines, polylines, splines, and elliptical arcs, and they must form a closed loop and share endpoints. Coons patch is a bicubic surface (one curve in the *M* direction and another in the *N* direction) interpolated between the four edges.

```
Command: EDGESURF ↵
Current wire frame density:  SURFTAB1=6  SURFTAB2=6
Select object 1 for surface edge: Select 1st object
Select object 2 for surface edge: Select 2nd object
Select object 3 for surface edge: Select 3rd object
Select object 4 for surface edge: Select 4th object
```

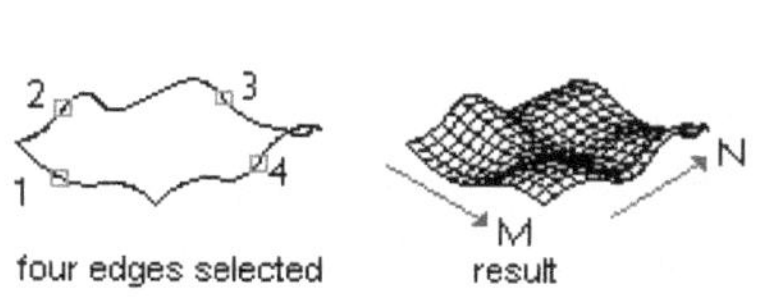

Figure 5.7 Edge-defined surface

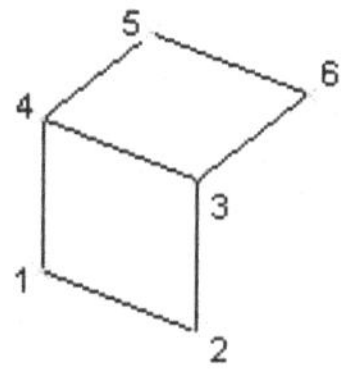

Figure 5.8 3D face

3DFACE Command

Pull-down menu: Draw → Modeling → Meshes → 3D Face
Command: 3DFACE ↵

3DFACE creates a three- or four-sided surface anywhere in 3D space (Figure 5.8). You can specify different Z coordinate values for each corner point of a 3D face, but if you do, the 3D face cannot be extruded. 3DFACE creates a surface that is not filled in; SOLID creates a filled-in surface.

```
Command: 3DFACE ↵
Specify first point or [Invisible]:  Specify a point (1)
Specify second point or [Invisible]:  Specify a point (2)
Specify third point or [Invisible] <exit>:  Specify a point (3)
Specify fourth point or [Invisible] <create three-sided face>:
                                                    Specify a point (4)
```

AutoCAD repeats the *third point* and *fourth point* prompts until you press NULL ENTER. Specify points 5 and 6 at these repeating prompts. When you finish entering points, press NULL ENTER.

Solid Modelling

A solid object represents the entire volume of an object. Solids are the most complete and least 3D modelling. Complex solid shapes are also easier to construct and edit than wireframes and meshes. You can create solids from one of the basic solid shapes, such as, box, cone, cylinder, sphere, torus, and wedge or by extruding a 2D object along a path or revolving a 2D object about an axis.

Once you create a solid in this way, you can create more complex shapes by combining these basic solid objects. You can join solids, subtract solids from one other, or find the common volume (overlapping portion) of solids. Solids can be further modified by filleting, chamfering, or changing the color of their edges. Faces on solids are easily manipulated because they do not require you to draw any new geometry or perform Boolean operations on the solid. AutoCAD also provides commands for slicing a solid into two pieces or obtaining the 2D cross section of a solid object.

Solid Box

Modeling toolbar: Box
Pull-down menu: Draw → Modeling → Box
Command: BOX ↵

You can use the BOX command to create a solid box. The base of the box is always parallel to the XY plane of the current UCS.

Corner option

```
Command: BOX ↵
Specify first corner or [Center] <0,0,0>: Specify corner
Specify corner or [Cube/Length]: Specify in relative coordinates
Specify height or [2Point]: Specify height of the box
```

Center-length option

```
Command: BOX ↵
Specify first corner or [Center] <0,0,0>: C ↵
Specify center < 0,0,0>: Specify center of the box
Specify corner or [Cube/Length]: L ↵
Specify length: Specify length
Specify width: Specify width
Specify height or [2Point]: Specify height
```

Solid Cone

Modeling toolbar: Cone
Pull-down menu: Draw → Modeling → Cone
Command: CONE ↵

You can use the CONE command to create a solid cone defined by a circular or an elliptical base tapering to a point perpendicular to its base. By default, the cone's base lies on the XY plane of the current UCS. The height, which can be positive or negative, is parallel to the Z axis. The apex determines the height and orientation of the cone.

```
Command: CONE ↵
Specify center point of base or [3P/2P/Ttr/Elliptical]:
                                              Specify center point
Specify base radius or [Diameter]: Specify radius
Specify height or [2Point/Axis endpoint/Top radius]: Specify height
```

Solid elliptical cones can also be created similarly.

Solid Cylinder

Modeling toolbar: Cylinder
Pull-down menu: Draw → Modeling → Cylinder
Command: CYLINDER ↵

You can use the CYLINDER to create a solid cylinder with a circular or an elliptical base. The base of the cylinder lies on the XY plane of the current UCS.

```
Command: CYLINDER ↵
Specify center point of base or [3P/2P/Ttr/Elliptical]: Specify center point
Specify base radius or [Diameter]: Specify radius
Specify height or [2Point/Axis endpoint]: Specify height
```

Solid elliptical cylinders can also be created similarly.

Solid Sphere

Modeling toolbar: Sphere
Pull-down menu: Draw → Modeling → Sphere
Command: SPHERE ↵

The SPHERE command creates a sphere with user defined center point and radius/diameter.

```
Command: SPHERE ↵
Specify center point or [3P/2P/Ttr]: Specify center point
Specify radius or [Diameter]: Specify radius or D for diameter
```

Solid Torus

Modeling toolbar: Torus
Pull-down menu: Draw → Modeling → Torus
Command: TORUS ↵

You can use the TORUS command to create a ring-shaped solid similar to the inner tube of a tire. The torus is parallel to and bisected by the XY plane of the current UCS. A torus may be self-intersecting. A self-intersecting torus has no center hole because the radius of the tube is greater than the radius of the torus.

```
Command: TORUS ↵
Specify center point or [3P/2P/Ttr]: Specify center point
Specify radius or [Diameter]: Specify radius
Specify tube radius or [2Point/Diameter]: Specify radius
```

Solid Wedge

Modeling toolbar: Wedge
Pull-down menu: Draw → Modeling → Wedge
Command: WEDGE ↵

You can use the WEDGE command to create a solid wedge. The base of the wedge is parallel to the XY plane of the current UCS with the sloped face opposite the first corner. Its height, which can be positive or negative, is parallel to the Z axis. Figure 5.9 shows a wedge of height 3 unit.

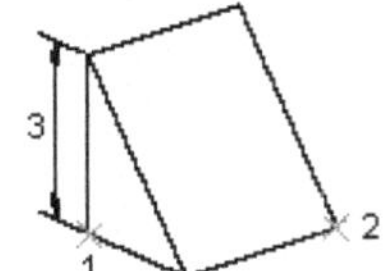

Figure 5.9 One example of wedge

```
Command: WEDGE ↵
Specify first corner or [Center]: Specify 1st corner point
Specify other corner or [Cube/Length]: Specify 2nd corner point
Specify height or [2Point]: Specify height
```

EXTRUDE Command

Modeling toolbar: Extrude
Pull-down menu: Draw → Modeling → Extrude
Command: EXTRUDE ↵

With the EXTRUDE command you can create solids by extruding (adding thickness to) selected 2D objects (Figure 5.10). You can extrude closed objects such as polylines, polygons, rectangles, circles, ellipses, closed splines, donuts, and regions. You cannot extrude 3D objects, objects contained within a block, polylines that have crossing or intersecting segments, or polylines that are not closed. You can extrude an object along a path, or you can specify a height value and a tapered angle.

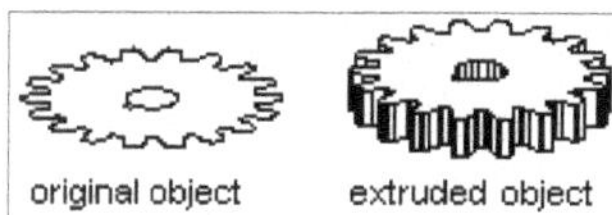

Figure 5.10 Extruding an object

```
Command: EXTRUDE ↵
Current wire frame density:  ISOLINES=4,
Closed profiles creation mode = Solid
Select objects to extrude or [MOde]: Select the object
Select objects to extrude or [MOde]: ↵
Specify height of extrusion or [Direction/Path/Taper angle/
Expression]: Specify the height
```

Tapering the extrusion is useful specifically for parts that need their sides defined along an angle, such as a mold used to create metal products in a foundry. Avoid using extremely large tapered angles. If the angle is too large, the profile can taper to a point before it reaches the specified height.

REVOLVE Command

Modeling toolbar: Revolve
Pull-down menu: Draw → Modeling → Revolve
Command: REVOLVE ↵

With the REVOLVE command, you can create a solid by revolving a closed object about the *X* or *Y* axis of the current UCS by using a specified angle. You can also revolve the object about a line, polyline, or two specified points (Figure 5.11). Similar to EXTRUDE, REVOLVE is useful for objects that contain fillets or other details that would otherwise be difficult to reproduce in a common profile. If you create a profile using lines or arcs that meet a polyline, use the PEDIT command and *join* option to convert them into a single polyline object before you use REVOLVE.

You can use REVOLVE on closed objects such as polylines, polygons, rectangles, circles, ellipses, and regions. You cannot use REVOLVE on 3D objects, objects contained within a block, polylines that have crossing or intersecting segments, or polylines that are not closed.

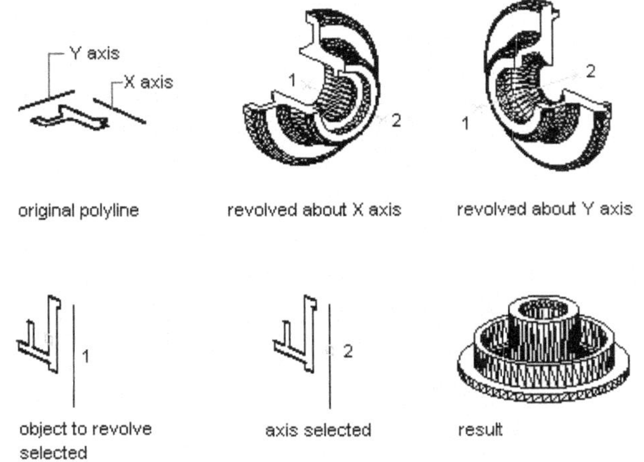

Figure 5.11 Using REVOLVE command

```
Command: REVOLVE ↵
Current wire frame density:  ISOLINES=4,
Closed profiles creation mode = Solid
Select objects to revolve or [MOde]: Select the object
Select objects to revolve or [MOde]: ↵
Specify axis start point or define axis by [Object/
X/Y/Z] <Object>: Specify the axis
Specify angle of revolution or [STart angle/Reverse/
EXpression] <360>: Specify the angle
```

Creating a Composite Solid

You can combine, subtract, and find the intersection of existing solids to create composite solids.

UNION Command

Solid Editing toolbar and Modeling toolbar: Union

Pull-down menu: Modify → Solids Editing → Union

Command: UNION ↵

With the UNION command, you can combine the total volume of two or more solids or two or more regions into a composite object (Figure 5.12).

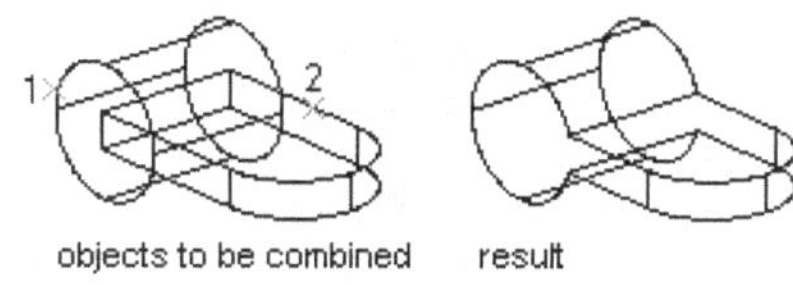

Figure 5.12 Using UNION command

SUBTRACT Command

Solid Editing toolbar and Modeling toolbar: Subtract
Pull-down menu: Modify → Solids Editing → Subtract
Command: SUBTRACT ↵

With SUBTRACT, you can remove the common area of one set of solids from another. For example, you can use SUBTRACT to add holes to a mechanical part by subtracting cylinders from the object (Figure 5.13).

Figure 5.13 Using SUBTRACT command

Figure 5.14: Using INTERSECT command

INTERSECT Command

Solid Editing toolbar and Modeling toolbar: Intersect
Pull-down menu: Modify → Solids Editing → Intersect
Command: INTERSECT ↵

With the INTERSECT command, you can create a composite solid from the common volume of two or more overlapping solids. INTERSECT removes the non-overlapping portions and creates a composite solid from the common volume (Figure 5.14).

The INTERFERE command performs the same operation as INTERSECT, but INTERFERE keeps the original two objects.

3D Array

Pull-down menu: Modify → 3D Operation → 3D Array
Command: 3DARRAY ↵

The command sequence is similar to 2D array, only an extra dimension comes into play in case of 3D array, i.e., the number of levels along Z-axis (Figure 5.15).

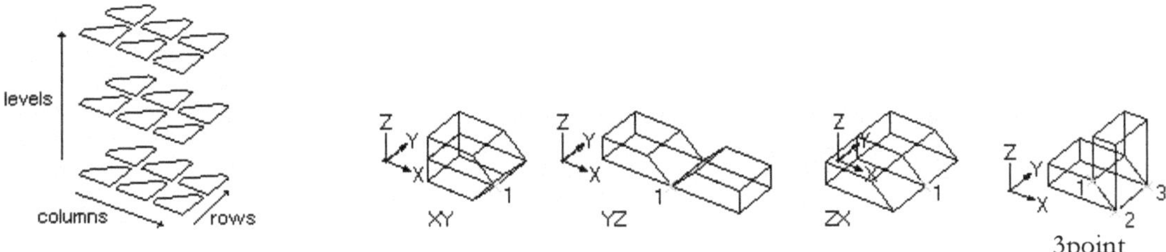

Figure 5.15 3D Array

Figure 5.16 Using MIRROR3D command

Mirroring 3D Objects

Pull-down menu: Modify → 3D Operation → 3D Mirror
Command: MIRROR3D ↵

The MIRROR3D command lets you mirror objects along a specified mirroring plane. The mirroring plane can be one of the following (Figure 5.16):

The plane of a planar object
A plane parallel to the XY, YZ, or XZ plane of the current UCS that passes through a selected point
A plane defined by three points that you select

```
Command: MIRROR3D ↵
Select objects: Select objects
Specify first point of mirror plane (3 points) [Object/Last/
Z axis/View/XY/YZ/ZX/3points] <3points>: Specify 3 points or an option
```

Rotating 3D Objects

Pull-down menu: Modify → 3D operation → 3D Rotate
Command: ROTATE3D or 3DROTATE ↵

The ROTATE3D command rotates objects in 3D about a specified axis. You can specify the axis of rotation by using two points; an object; the X, Y, or Z axis; or the Z direction of the current view. To rotate 3D objects, you can use either ROTATE3D or 3DROTATE. The ROTATE command can also be used to rotate a 3D object on XY plane.

Command: **3DROTATE ↵**
Select objects: *Select the 3D object*
Select objects: ↵
Specify base point: *Define the base point*
Specify angle start point or type an angle: *Specify a point*
Specify angle end point: *Specify another point*

Filleting 3D Solids

Modify toolbar: Fillet
Pull-down menu: Modify → Fillet
Command: FILLET or F ↵

With the FILLET command, you can add rounds and fillets to selected objects. The default method is to specify the fillet radius and then selecting the edges to fillet (Figure 5.17). Other methods specify individual measurements for each filleted edge and fillet a tangential series of edges.

Command: **FILLET ↵**
Current settings: Mode = TRIM, Radius = 0.0000
Select first object or [Undo/Polyline/Radius/Trim/Multiple]: *Select the object*
Enter fillet radius or [Expression]: *Enter fillet radius*
Select an edge or [Chain/Radius]: *Select the edge*
Select an edge or [Chain/Radius]: *Select the edge*
Select an edge or [Chain/Radius]: *Select the edge*

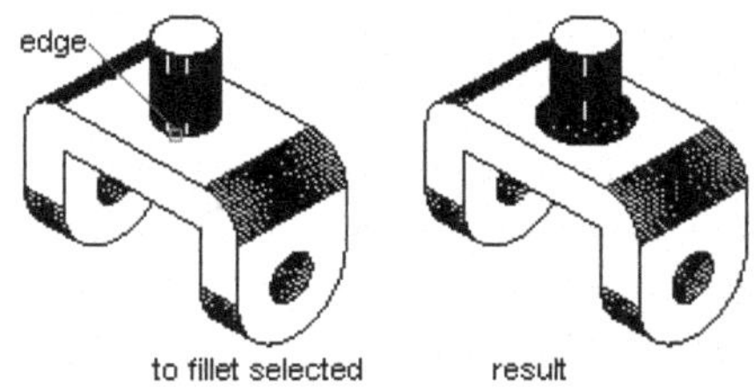

Figure 5.17 Using FILLET command

Chamfering 3D Solids

Modify toolbar: Chamfer
Pull-down menu: Modify → Chamfer
Command: CHAMFER or CHA ↵

The CHAMFER command bevels the edges along the adjoining faces of a solid (Figure 5.18).

Command: **CHAMFER ↵**
(TRIM mode) Current chamfer Dist1 = 0.0000, Dist2 = 0.0000
Select first line or [Undo/Polyline/Distance/Angle/Trim/mEthod/Multiple]: *Select the object (1 in Figure 5.18)*
Base surface selection...
Enter surface selection option [Next/OK (current)] <OK>: ↵
Specify base surface chamfer distance: *Enter the distance*
Specify other surface chamfer distance <default>: *Enter the distance*
Select an edge or [Loop]: *Select the edge (2 in Figure 5.18)*

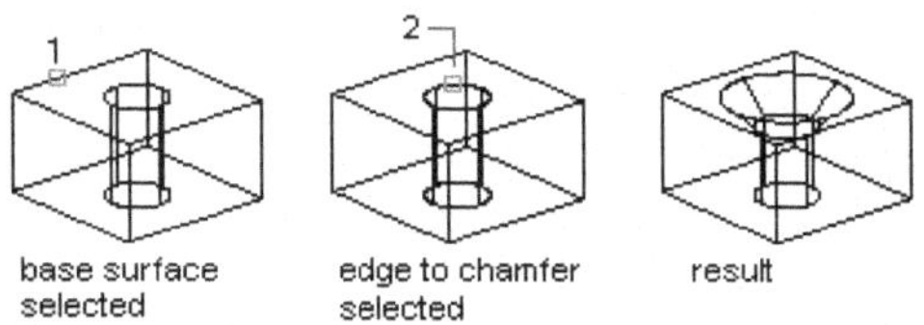

Figure 5.18 Using CHAMFER command

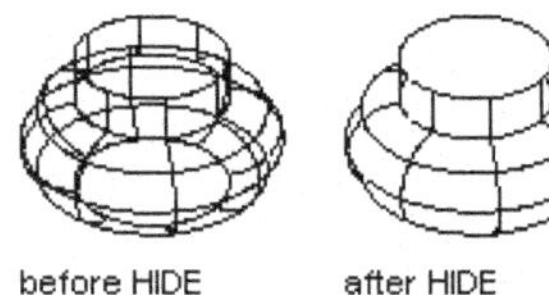

Figure 5.19 HIDE command

HIDE Command

Render toolbar: Hide
Pull-down menu: View → Hide
Command: HIDE or HI ↵

The HIDE command regenerates a three-dimensional model with hidden lines suppressed (Fugure 5.19). When you draw 3D objects, AutoCAD produces a wireframe display in the current viewport where all lines are present, including those hidden by other objects. HIDE eliminates the hidden lines from the screen. REGEN command gets back the hidden lines again.

SHADEMODE Command

Command: SHADEMODE or SHA ↵

The SHADEMODE command provides shading and wireframe options for the objects in the current viewport. You can edit shaded objects without regenerating the drawing. Although hiding of lines enhances the drawing and clarifies the design, shading produces a more realistic image of your model. SHADEMODE provides you with options to view and edit your objects in wireframe or shaded representations.

```
Command: SHADEMODE ↵
Enter an option
[2dwireframe/Wireframe/Hidden/Realistic/Conceptual/Shaded/
shaded with Edges/shades of Gray/SKetchy/X-ray/
Other] <2dwireframe>: Enter an option
```

Widely used options are as follows:

2D Wireframe: Displays the objects by using lines and curves to represent the boundaries. Raster and OLE objects, linetypes, and lineweights are visible. Even if the value for the COMPASS system variable is set to 1, it does not appear in the 2D Wireframe view.

3D Wireframe: Displays the objects by using lines and curves to represent the boundaries. Displays a shaded 3D UCS icon. Raster and OLE objects, linetypes, and lineweights are not visible. You can set the COMPASS system variable to 1 to view the compass. Material colors which you have applied to the objects are shown.

3D Hidden: Displays the objects using 3D wireframe representation and hides lines representing back faces.

Realistic: Shades the objects and smooths the edges between polygon faces. Materials that you have attached to the objects are displayed.

Conceptual: Shades the objects and smooths the edges between polygon faces. Shading uses a transition between cool and warm colors. The effect is less realistic, but it can make the details of the model easier to see.

Other: Displays the following prompt:

```
Enter a visual style name [?]: Enter the name of a visual style in the
        current drawing, or enter ? to display a list of names and repeat the prompt
```

RENDER Command

Render toolbar: Render

Pull-down menu: View → Render → Render

Command: RENDER or RR ↵

RENDER produces an image by using information from a scene, the current selection set, or the current view. It creates a photorealistic or realistically shaded image of a three-dimensional wireframe or solid model (Figure 5.20). AutoCAD uses geometry, lighting, and materials to render a realistic image of a model. For a presentation, a full rendering might be appropriate.

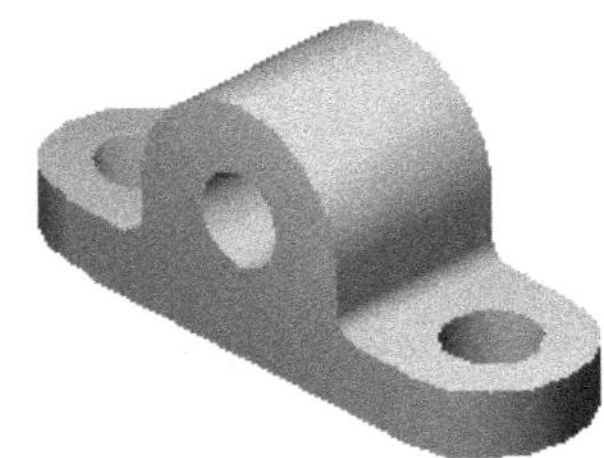

Figure 5.20 A rendered object

RMAT Command

Render toolbar: Materials
Pull-down menu: View → Render → Materials Browser
Command: RMAT ↵

To lend still greater realism to your renderings, apply materials such as steel and plastic to the surfaces of your model. You can attach materials to individual objects, all objects with a specific AutoCAD Color Index (ACI) number, blocks, or layers.

Using materials involves several steps:

- Defining materials, including their color, reflection, or dullness
- Attaching materials to objects in the drawing
- Importing and exporting materials to and from material libraries
- Creating color, shading, and patterning is different on a computer than with traditional media such as paints and crayons.

LIGHT Command

Render toolbar: Lights
Pull-down menu: View → Render → Light → Point Light / Spot Light / …
Command: LIGHT ↵

The LIGHT command manages lights and lighting effects. You can use light in model space only.

DVIEW Command

Command: DVIEW or DV ↵

The DVIEW command defines parallel projection or perspective views. To help you viewing a model from any point in space, DVIEW uses a camera-target metaphor. The line of sight, or viewing direction, is the line between the camera and the target.

`Command:` **DVIEW ↵**
`Select objects or <use DVIEWBLOCK>:` Enter option
`[CAmera/TArget/Distance/POints/PAn/Zoom/TWist/CLip/`
`Hide/Off/Undo]:` *Specify a point with your pointing device, or enter an option*

PLAN Command

Pull-down menu: View → 3D Views → Plan View → Current UCS
Command: PLAN ↵

This command displays the plan view of an object. PLAN command provides a convenient means of viewing the drawing from the plan view. You can select a plan view of the current user coordinate system, a previously saved UCS, or the world coordinate system. PLAN affects the view in the current viewport only. You cannot use PLAN command in paper space.

The PLAN command changes the viewing direction and turns off perspective and clipping; it does not change the current UCS. Any coordinates entered or displayed subsequent to the PLAN command remains relative to the current UCS.

`Command:` **PLAN ↵**
`Enter an option [Current ucs/Ucs/World] <Current>:` *Enter an option*

MASSPROP Command

Inquiry toolbar: Region/Mass Properties
Pulldown menu: Tools → Inquiry → Region/Mass Properties
Command: MASSPROP ↵

The MASSPROP command calculates properties of 2D and 3D objects which are essential in analyzing the characteristics of the drawn objects. It displays the mass properties in the text window, and then asks if you want to write the mass properties to a text file.

The properties that MASSPROP command displays depend on whether the selected objects are regions, and whether the selected regions are coplanar with the XY plane of the current user coordinate system (UCS), or 3D solids.

The following table shows the parameters that control the units in which mass properties are calculated.

Parameter	*Used to calculate*
DENSITY	Mass of solids
LENGTH	Volume of solids
LENGTH*LENGTH	Area of regions and surface area of solids
LENGTH*LENGTH*LENGTH	Bounding box, radii of gyration, centroid, and perimeter
DENSITY*LENGTH*LENGTH	Moments of inertia, products of inertia, and principal moments

6
AutoLISP

Introduction

AutoLISP is an implementation of the LISP programming language which comes integrated with AutoCAD. This facility gives an opportunity to programmatically use the features of AutoCAD. AutoLISP is a subset of common LISP, the most recent version of the oldest artificial intelligence programming language still in use today. AutoLISP is essentially a pared down version of Common LISP with some additional features unique to AutoCAD.

Data Type

Following are the data types accepted in AutoLISP:
Lists (the most important types), symbols, strings, real nos., integers, file descriptors, AutoCAD entity names, subrs, atoms etc.

Integers may range from –32768 to +32767. Ex: Integer = 24
Real numbers are represented as double precision floating points. Ex: Real Number = 0.618
Strings can be of any length. String constants can be of a maximum of 132 characters.
Ex: 'Enter the temp. in centigrade:'
Subrs are AutoLISP built-in functions. Ex: Subrs

Lists are group of related values separated by space and enclosed in a pair of parentheses (Figure 6.1).

Ex.: `(a b c)`

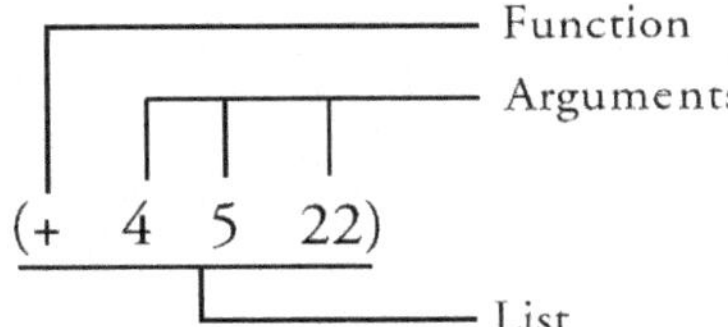

Figure 6.1 Function and arguments of a list

AutoLISP allows you to read and write text files to disk. File descriptors are used in a program to access files that have been opened for processing.

There are two classes of data, *lists* and *atoms*. An atom is an element that cannot be taken apart into other elements. For example, a coordinate list can be 'disassembled' into three numbers, the *X* value, the *Y* value, and the *Z* value, but these values cannot be taken apart any further. In a coordinate list, the *X*, *Y*, and *Z* values are atoms. Symbols are also atoms because they are treated as single object. So, in general, atoms are either numbers or symbols.

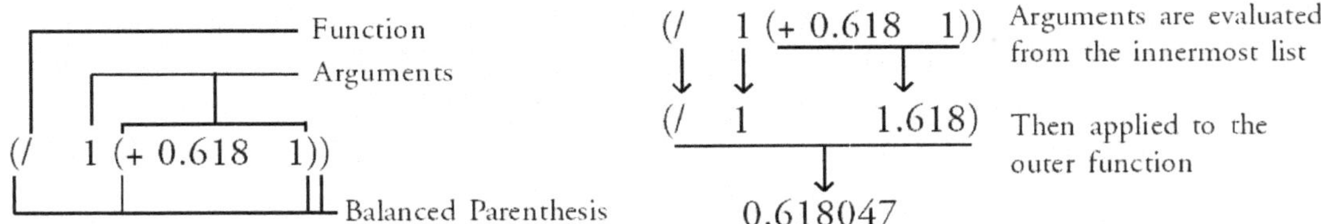

Figure 6.2 Items of a List and how a List works

Point representation

2D points are expressed as a list of two real numbers.
3D points are expressed as a list of three real numbers.

Order of elements in the list is important.

For a 2D point, the order is (x y).
For a 3D point, the order is (x y z).

AutoLISP evaluation convention

i) Integers, real numbers, strings, etc. evaluate to themselves.
ii) Symbols evaluate to their current binding.
iii) Lists are evaluated according to the first element of me list as follows:
    If the first element is a function name, it is evaluated by using the remaining elements as arguments.
    If the first element is a subr, the subr is evaluated with the remaining elements as arguments.

Remember the following:

1. All the expressions in AutoLISP are evaluated to some value, or true (=T), or False (=nil).
2. A list without any element is an empty list, and is denoted as ().
3. '()=nil
4. All expressions in AutoLISP are evaluated. To stop evaluation, a quote (') is used.
   Ex: `'(+ 1 2)` stops addition of 1 and 2.

Naming convention

A variable name may begin with any character, but may not contain only numerals. It should also not contain the characters (), / " and ;

Ex: `ab#` is a correct variable name.
`123` is an incorrect variable name.

It is advisable not to use any special characters in the name other than _ (underscore), and to always begin with an alphabet.

Ex: `ab1`

Program structure

All the AutoLISP statement forms are as follows:
(<function name> <list of arguments>) The function name and the arguments should all be separated by a single blank space.

Ex:
```
(+ 1 2 3)
(max 3 29 31 20)
```

Arithmetical functions

1. **(+ <list of nos.>)**
   This adds all the nos. and returns the result.
   Ex: `(+ 2 3)` Returns: `5`
   `(+ -2 -3)` Returns: `-5`

2. **(– <list of nos.>)**
   This subtracts the sum of all the nos. from the second no. to the last, from the first no. and returns the result.
   Ex: `(- 10 4 1)` Returns: `5`

3. **(* <list of nos.>)**
   This multiplies all the nos. and returns the result.
   Ex: `(* 3 4)` Returns: `12`

4. **(/ <list of nos.>)**
   This divides the first no. from the product of the second no. through last no, and returns the result.
   Ex: `(/ 8 4 )` Returns: `2`
   `(/ 8 4 2)` Returns: `1`

5. **(1+ <no>)**
   This function returns the argument increased by one.
   Ex: `(1+ 5)` Returns: `6`

6. **(1– <no>)**
   This function returns the argument decreased by one.
   Ex: `(1- 5)` Returns: `4`

7. **(abs <no>)**
   This function returns the absolute value of the no.
   Ex: `(abs 100)` Returns: `100`
   `(abs -100)` Returns: `100`

8. **(fix <no>)**
   This returns an integer value depending on the <no>. If <no> is an integer, then the integer is returned, otherwise the no. is returned with the decimal part truncated.
   Ex: `(fix 1)` Returns: `1`
   `(fix 1.9)` Returns: `1`

9. **(float <no>)**
This returns a real no. depending on the <no>. If <no> is an integer, then it is returned with a decimal part, otherwise the <no> is returned.
Ex: `(float 3)` Returns: `3.000000`
`(float 3.7)` Returns: `3.700000`

10. **(log <no>)**
This returns the e based log of <no>.
Ex: `(log 4.5)` Returns: `1.504077`

11. **(max <list of nos.>)**
This returns the maximum no. in the <list of nos.> taking sign into account.
Ex: `(max 1 7 -3)` Returns: `7`

12. **(min <list of nos.>)**
This returns the minimum no. in the <list of nos.> taking sign into account.
Ex: `(min 1 -2 33 2)` Returns: `-2`

13. **(sqrt <no>)**
This returns the square root of the <no>.
Ex: `(sqrt 4)` Returns: `2.000000`

Logical functions

1. **(and <expr>...)**
This returns the logical AND of the <expr> list.
Ex: `(and (= 4 4) (> 3 6))` Returns: `nil`
`(and (= 4 4) (< 3 6))` Returns: `T`

2. **(or <expr>...)**
This returns the logical OR of the <expr> list.
Ex: `(or (= 4 4) (> 3 6))` Returns: `T`
`(or (= 3 4) (> 3 6))` Returns: `nil`

3. **(not <expr>...)**
This returns T if the item is false and returns FALSE if it is true.
Ex: `(not (< 3 2))` Returns: `T`
`(not (> 3 2))` Returns: `nil`

Geometric functions

1. **(angle <ptl> <pt2>)**
   This determines the angle between the points <ptl> and <pt2> in radians.
   Ex: `(angle '(1.0 1.0) '(1.0 4.0))` Returns: `1.570796`

2. **(cos <angle>)**
   This determines the cosine of the <angle> given in radians.
   Ex: `(cos 0.0)` Returns: `1.00000`

3. **(sin <angle>)**
   This determines the sine of the <angle> given in radians.
   Ex: `(sin 1.0)` Returns: `0.841471`

4. **(distance <ptl> <pt2>)**
   This determines the distance between two points on the screen.
   Ex: `(distance '(1.0 2.5) '(7.7 2.5))` Returns: `6.700000`

5. **(polar <pt> <angle> <distance>)**
   This function returns the point at an angle <angle>, distance <distance>, from point <pt>.
   Ex: `(polar '(1.0 1.0) 0.785 1.414)`
   Returns: `(2.000000 2.000000)`
   Remember, the <angle> must be in radians.

Arithmatic Operators

1. **(= <atom> <atom> ...)**
   This returns T if all the specified atoms are numerically equal. Otherwise it returns nil.
   Ex: `(= 4 4)` Returns: `T`
   `(= 4 3)` Retuns: `nil`

2. **(/= <atom> <atom>...)**
   This returns T if all the specified atom are numerically unequal. Otherwise it returns nil.
   Ex: `(/= 4 4)` Returns: `nil`
   `(/= 4 5)` Returns: `T`

3. **(< <atom> <atom>...)**
   This returns T if the atoms are numerically lesser than the succeeding ones.
   Ex: `(< 1 2 3)` Retuns: `T` and `(< 4 3)` Returns: `nil`

4. **(<= <atom> <atom>...)**
This returns T if all the atom are numerically lesser than or equal to the succeeding ones. Otherwise it returns nil.
Ex: `(<= 10 4)` Returns: `nil`
`(<= 10 10)` Returns: `T`

5. **(> <atom> <atom>...)**
This returns T if all the atoms are numerically greater than the succeeding ones. Otherwise it returns nil.
Ex: `(> 11 10)` Returns: `T`
`(> 10 11)` Returns: `nil`

6. **(>= <atom> <atom>...)**
This returns T if all the atoms are numerically greater than or equal to the succeeding ones. Otherwise it returns nil.
Ex: `(>= 11 10)` Returns: `T`
`(>= 10 11)` Returns: `nil`

List Functions

1. **(append <expr>...)**
This takes any no. of lists and forms a single list from their elements.
Ex: `(append '(a b) '(x y))` Returns: `(a b x y)`

2. **(car <list>)**
This returns the first element of the <list>.
Ex: `(car '(a b))` Returns: `a`

3. **(cdr <list>)**
This function returns the given <list> without the first element.
Ex: `(cdr '(a b))` Returns: `(b)`
AutoLISP support concatenation of car and cdr up to four levels deep.
Ex: `(car (car '((a b) c)))` may be represented as: `(caar '((a b) c))`
Returns: `a`
`(car (cdr '(a b c)))` may be represented as: `(cadr '(a b c))`
Returns: `b`

4. **(last <list>)**
This returns the last element of the <list>.
Ex: `(last '(a b c))` Returns: `c`

5. **(length <list>)**
This returns the integer indicating the no. of elements in <list>.
Ex: `(length (a b c))` Returns: `3`

6. **(list <expr>...)**
This function takes any no. of arguments and strings them together to form a new list.
Ex: `(list '(a b) '(c d))` Returns: `((a b) (c d))`

7. **(progn <expr>...)**
This evaluates each <expr>, and returns the value of the last one.

8. **(reverse <list>)**
This function reverses the order of the elements in the <list>.
Ex: `(reverse '(a b c))` Returns: `(c b a)`

AutoCAD Command

1. **(command <args>...)**
This function executes AutoCAD commands directly from AutoLISP.
Ex: `(command "line" pt1 pt2 "")`
This will draw a line from the pt1 to pt2.
The symbol `""` denotes null return (null ENTER).

Conditional Functions

1. **(cond (<test-1> <result-1>)**
(<test-2> <result-2>)
....
(<test-n> <result-n>))
This function evaluates each <test-n> until it finds one which is T. Then it returns the corresponding <result-n>. If no T <test-n> is found, nil is returned.

2. **(if <cond> <then-body> [<else-body>])**
This evaluates the <then-body> if <cond> is true. Otherwise it executes the <else-body> if it is present. You can use this function without <else-body>.

Loop Functions

1. **(while <test> <body>)**
   This evaluates the <test> clause, and as long as it is T, <body> is evaluated.

2. **(repeat <no> <expr>...)**
   This evaluates <expr> <no> of times, and returns the value of the last expression.

Assingment Functions

1. **(setq <symbol> <value>...)**
   This function assigns values to the preceding symbols. Any number of symbols can be assigned values.
   Ex: `(setq a 3 b 5)` assigns the value `3` to `a` and the value `5` to `b`.

2. **(setvar <var> <val>)**
   This sets the value of an AutoCAD system variable.
   Ex: `(setvar "blipmode" 0)` sets the value of blipmode to `0`.

Input (Get) Functions

1. **(getangle [<pt>] [<prompt>])**
   This function pauses for the input of two points and returns the angle between them in radians. <pt> is an optional 2D point, and <prompt> an optional string to be displayed.
   Ex: `(getangle '(1.0 3.5) "The other point:")`
   This waits for the user input of a point with the string "`The other point:`" as the prompt.

2. **(getdist [<pt>] [<prompt>])**
   This function pauses for the input of two points and returns the distance between them. <pt> is an optional 2D point, and <prompt> is an optional string to be displayed.
   Ex: `(getdist '(1.0 3.5) "The other point:")`
   This waits for the user input of a point with the string "`The other point:`" as the prompt.

3. **(getpoint [<prompt>])**
   This function pauses for the input of a point and returns a point. <prompt> is an optional string to be displayed.

Ex: `(getpoint "Enter the point:")`

This waits for the user input of a point with the string "`Enter the point:`" as the prompt.

4. **(getint [<prompt>])**

   This function pauses for user input of a integer number (without having decimal part), and returns that integer number. <prompt> is an optional string to be displayed as a prompt.

   Ex: `(getint "Enter the number of circles:")`

   This waits for the user input of an integer number with the string "`Enter the number of circles:`" as the prompt.

5. **(getreal [<prompt>])**

   This function pauses for user input of a real number, and returns that real number <prompt> is an optional string to be displayed as a prompt.

   Ex: `(getreal "Enter the radius:")`

   This waits for the user input of a real number with the string "`Enter the radius:`" as the prompt.

6. **(getvar <var>)**

   This returns the value of the AutoCAD system variable <var>.

   Ex: `(getvar "filletrad")` Returns: `0.500000` (if it is set to that value).

7. **(initget [bits] [string])**

   The functions which honor key words are getint, getreal, getdist, getangle, getorient, getpoint, getcorner, getkword. The getstring function is the only user-input function which does not honor key words.

Bit value	***Description***
1 (bit 0)	Prevents the user from responding to the request by entering only ENTER.
2 (bit 1)	Prevents the user from responding to the request by entering zero.
4 (bit 2)	Prevents from responding to the request by entering a negative value.

\n within prompts generates a new line on the screen.

Ex: `"\nEnter the center point"`

Output Functions

1. **(prompt <msg>)**
   This displays the <msg> on the promt area.
   Ex: `(prompt "Enter point:")` Returns : `Enter point:`

2. **(print <expr>)**
   This prints an expression to the command line. This function prints a newline character before <expr>, and prints a space following <expr>.

3. **(princ <expr>)**
   This function prints the control characters in <expr> without expansion.

Function Definition

1. **(defun <name> <arguments> <body>)**
   This defines a function with the name <name> with <arguments> as its parameter list, and <body> as the expression(s) defining the function.

Graphic functions

1. **(graphscr)**
   This switchs the text screen to the graphics screen.

2. **(textscr)**
   This switches the graphics screen to the text screen.

3. **(redraw)**
   This redraws the entire drawing.

How to write and run programs

1. Open Visual LISP Editor form **Tools → AutoLISP → Visual LISP Editor**
2. Then the editing screen will appear. After the editing screen appears, create a new file and type the program lines.
3. Save the file.
4. To run the program, at first you have to load the program as follows: *(load "test")*, or click on *Tools* pull-down menu, then on *Load Application.....* A dialog box will

appear. Select file type as LSP. Select your LISP file and click on *Load.* Then close the appeared dialog box.

5. If loading is unsuccessful, it may be due to the typical errors like not matching parentheses, or not closing a quote mark etc.
6. To run the program, type the name of the first main function with it's arguments and press the ENTER key.
   For example if test.LSP file contains the main function 'cir1', just type *(cir1)* at command prompt like:
   `Command:` **(cir1)** ↵

Some typical error messages

1. **Bad argument type:** an incorrect type of argument was passed to a function.
2. **Null function:** an attempt was made to evaluate a function that has a nil definition.
3. **Bad list:** an improperly formed list was passed to a function.
4. **Divide by zero:** division by zero is not allowed.
5. **Bad point argument:** a poorly defined point was passed to a function expecting a point.
6. **Extra right paren:** one or more extra right parentheses were found.
7. **Invalid character:** an expression contains an improper character.
8. **Malformed list:** a list being read from a file which has ended prematurely.
9. **Malformed string:** a string being read from a file which has ended prematurely.
10. **Too few arguments:** too few arguments were passed to a built-in function.
11. **Too many arguments:** too many arguments were passed to a built-in function.

Some simple examples of AutoLISP program

1. Program for calculating the area of a circle. Input will be the radius of circle.

```
(defun test()
      (setq r (getreal "Enter the radius of the circle"))
      (setq a (* pi r r))
      (prompt "The area of the circle is:") (print a)
      (princ)
 )
```

2. Program for drawing a circle in AutoCAD. Inputs will be the center point and radius of the circle.

```
(defun cir1()
      (initget 1)
         (setq cp (getpoint "Enter the centerpoint of circle:"))
      (initget (+ 1 2 4))
         (setq r (getreal "Enter the radius of the circle:"))
      (command "circle" cp r)
 )
```

3. Program for drawing a rectangle by using LINE command. Inputs will be the lower-left point of the rectangle, width and length of the rectangle and the rotation angle.

```
(defun rec1()
       (initget 1)
        (setq p1 (getpoint "\nENTER THE LOWER-LEFT CORNER:"))
       (initget (+ 1 2 4))
        (setq w (getreal "\nENTER THE WIDTH:"))
       (initget (+ 1 2 4))
        (setq l (getreal "\nENTER THE LENGTH:"))
        (initget 1)
            (setq a (getreal "\nENTER THE ANGLE WITH X AXIS:"))
      (setq a (/ (* pi a) 180))
      (setq p2 (polar p1 a w))
           (setq p3 (polar p2 (+ (/ pi 2) a) l))
           (setq p4 (polar p3 (+ a pi) w))
     (command "line" p1 p2 p3 p4 "c")
 )
```

4. Program for drawing a circle and calculating its area. Inputs will be the center point and radius of the circle. Before drawing the circle the program will erase all objects from the file.

```
(defun mycir()
  (command "erase" "all" "")
  (setq cp (getpoint "\nEnter the center point Of the circle:"))
  (setq d1 (getreal "\nEnter the diameter:"))
  (setq r (/ d1 2))
  (setq a (* pi r r))
  (command "circle" cp "d" d1)
  (prompt "area of the circle is :")(print a)
  (princ)
  )
```

5. Program for drawing concentric circles using WHILE loop. Inputs will be the center point, radius of innermost circle, radial increment, and number of circles. Before drawing the circles the program will erase all objects from the screen.

```
(defun concir()
  (command "erase" "all" "")
  (initget 1)
  (setq cp (getpoint  "\nEnter the center point of the circle:"))
  (initget ( + 1 2 4))
  (setq r (getreal "\nEnter the radius of innermost circle:"))
  (initget (+ 1 2 ))
  (setq i (getreal "\nEnter inter radial increment:"))
  (if ( < i 0)(alert "You have entered -ve data !!!"))
  (initget (+ 1 2 4))
  (setq n (getint "Enter the no of the circles:"))
  (while (and (> n 0)(> r 0))
    (command "circle" cp r)
    (setq r (+ r i))
    (setq n (- n 1))
    )
  )
```

6. Program for drawing concentric circles using REPEAT loop. Inputs will be the center point, radius of innermost circle, radial increment, and number of circles. Before drawing the circles the program will erase all objects from the screen.

```
(defun c:concir()
  (command "erase"  "all"  "")
  (initget 1)
  (setq cp (getpoint "\nEnter the center point:"))
  (initget (+ 1 2 4))
  (setq r  (getreal "\nEnter the innermost radius:"))
  (initget (+ 1 2 ))
  (setq i (getreal "\nEnter the iner radial increment:"))
  (initget (+ 1 2 4 ))
  (setq n (getint "\nEnter the no of circles "))
  (if (< i 0)(alert "You have entered -ve data !!!"))
  (repeat n
    (command "circle" cp r)
    (setq r (+ r i))
    )
  )
```

7. Program for drawing a polygon using POLYGON command. Inputs will be the number of sides, center point, radius, and inscribed/circumscribed.

```
(defun poly1()
  (setq n (getint "\nEnter number sides:"))
  (setq cp (getpoint "\nEnter the center point:"))
  (setq s (getstring "Enter i for inscribed & c for circumscribed:"))
  (setq r (getreal "\nEnter the radius:"))
  (command "polygon" n cp s r)
  ;(command "circle" cp r)
  )
```

Now run the program again after removing the semicolon (;) of the last line, and notice the change in results.

8. Program for drawing parallel tilted rectangles. Inputs are as shown in Figure 6.3. Before drawing the rectangles the program will erase all objects from the file.

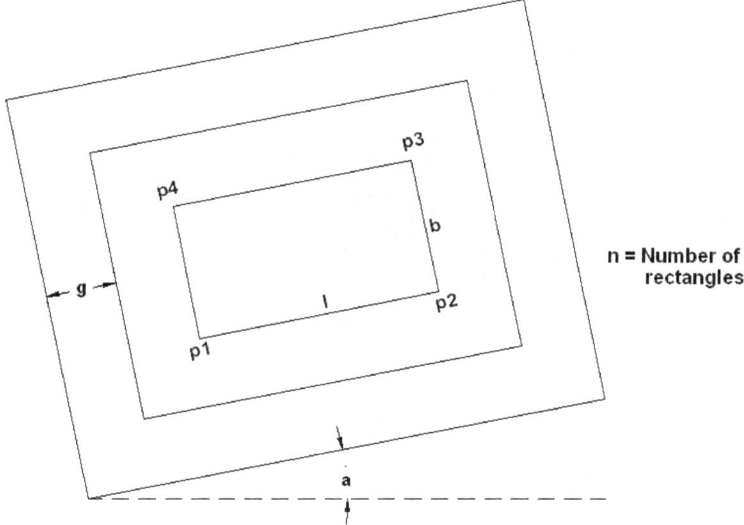

Figure 6.3 Inputs for parallel tilted rectangles

```
(defun c:conrec()
  (command "erase" "all" "")
  (setq p1 (getpoint "\nEnter the lower-left corner:"))
  (setq l (getreal "\nEnter the length:"))
  (setq b (getreal "\nEnter the breadth:"))
  (setq g (getreal "\nEnter inter rectangular distance:"))
```

```
(setq n (getint "\nEnter the no of rectangles:"))
(setq a (getreal "\nEnter the angles indegrees:"))
(setq a (/(* pi a)180))
(repeat n
  (setq p2 (polar p1 a l))
  (setq p3 (polar p2 (+ (/ pi 2)a)b))
  (setq p4 (polar p3 (+ pi a)l))
  (command "line" p1 p2 p3 p4 "c")
  (setq p1 (polar p1(+(/(* 5 pi)4)a)(* 1.414 g)))
  (setq l (+ l g g))
  (setq b (+ b g g))
  )
)
```

9. Program for drawing an isolated foundation. Inputs are as shown in Figure 6.4. Before drawing the foundation the program will erase all objects from the file.

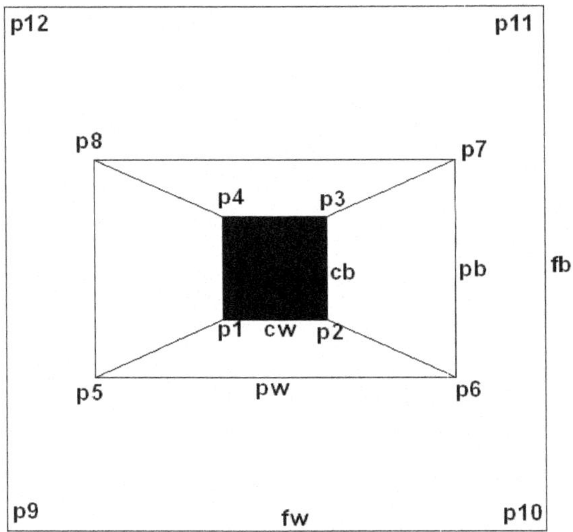

Figure 6.4 Inputs for isolated foundation

```
(defun c:footing()
  (command "erase" "all" "")
  (setq p1 (getpoint "\nEnter the lower-left corner of the column:"))
  (setq cw (getreal "\nEnter the width of the column:"))
  (setq cb (getreal "\nEnter the breadth of the column:"))
  (setq pw (getreal "\nEnter the width of the pedestal:"))
  (setq pb (getreal "\nEnter the breadth of the pedestal:"))
  (setq fw (getreal "\nEnter the width of the footing:"))
  (setq fb (getreal "\nEnter the breadth of the footing:"))
```

```
(setq p2 (polar p1 0 cw))
(setq p3 (polar p2 (/ pi 2) cb))
(setq p4 (polar p3 pi cw))

(setq p5 (list (-(car p1)(/(- pw cw)2))(-(cadr p1)(/(- pb cb)2))))
(setq p6 (polar p5 0 pw))
(setq p7 (polar p6 (/ pi 2) pb))
(setq p8 (polar p7 pi pw))

(setq p9 (list (- (car p5)(/(- fw pw)2))(-(cadr p5)(/(- fb pb)2))))
(setq p10 (polar p9 0 fw))
(setq p11(polar p10 (/ pi 2) fb))
(setq p12(polar p11 pi fw))

(command "solid" p1 p2 p4 p3 "")
(command "line" p5 p6 p7 p8 "c")
(command "line" p9 p10 p11 p12 "c")
(command "line" p1 p5 "")
(command "line" p2 p6 "")
(command "line" p3 p7 "")
(command "line" p4 p8 "")
)
```

Space for Notes

Appendix A
Exercises

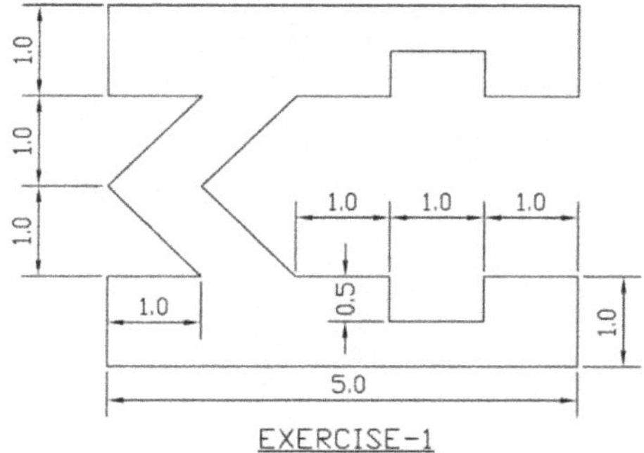

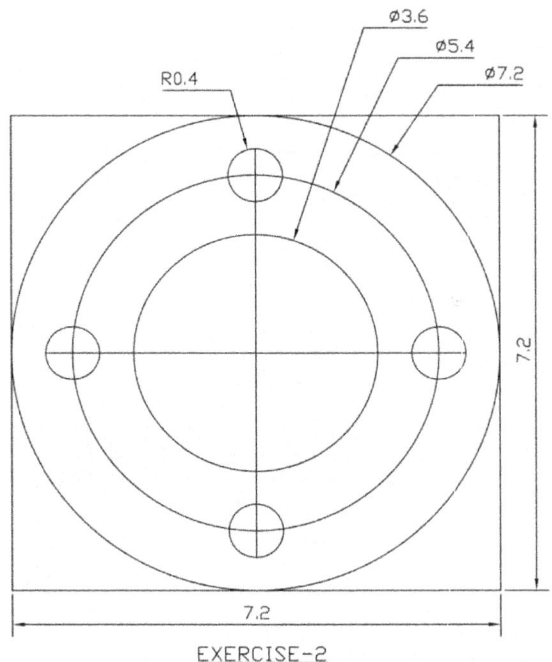

2.00 (ALL SIDES)

Ø4.00

R0.50 (TYP.)

0.50 (TYP.)

R0.50

Exercise-3

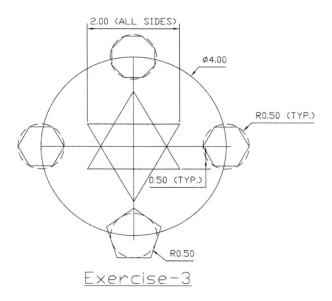

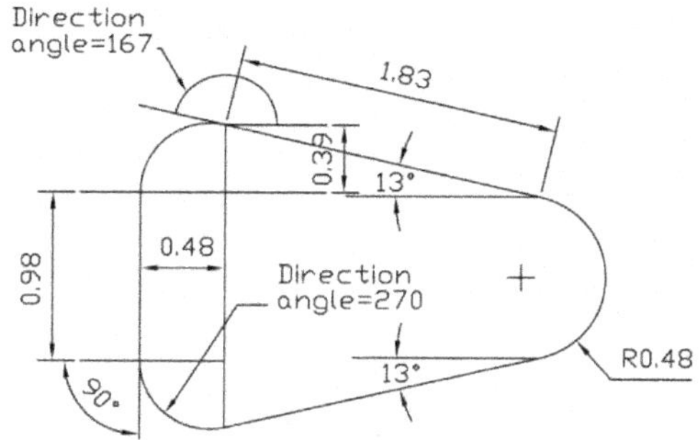

Exercise-4

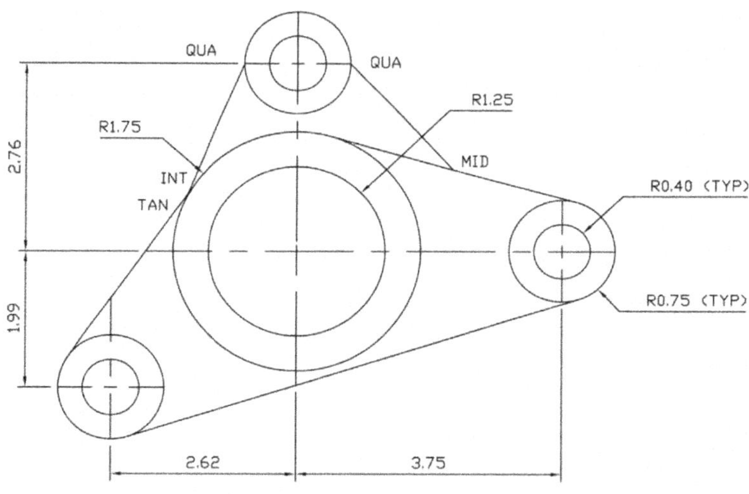

Exercise-5

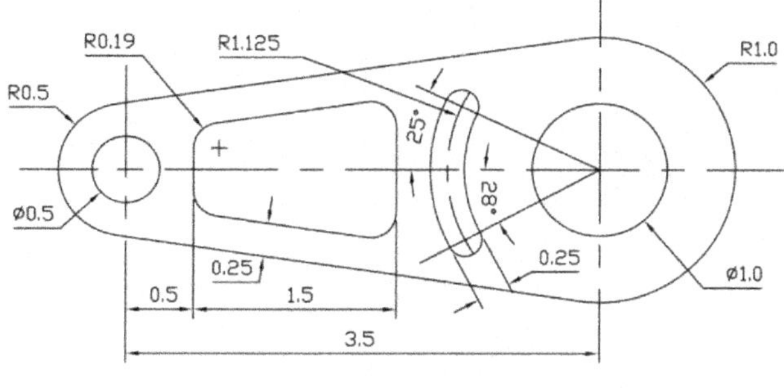

EXERCISE-6

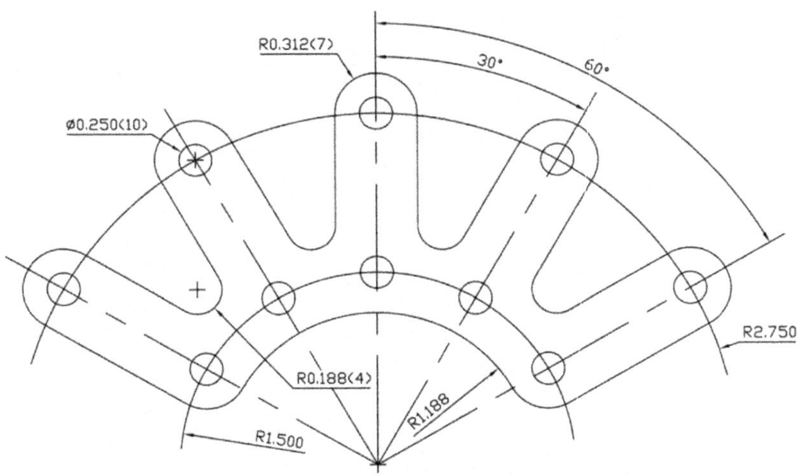

EXERCISE-7

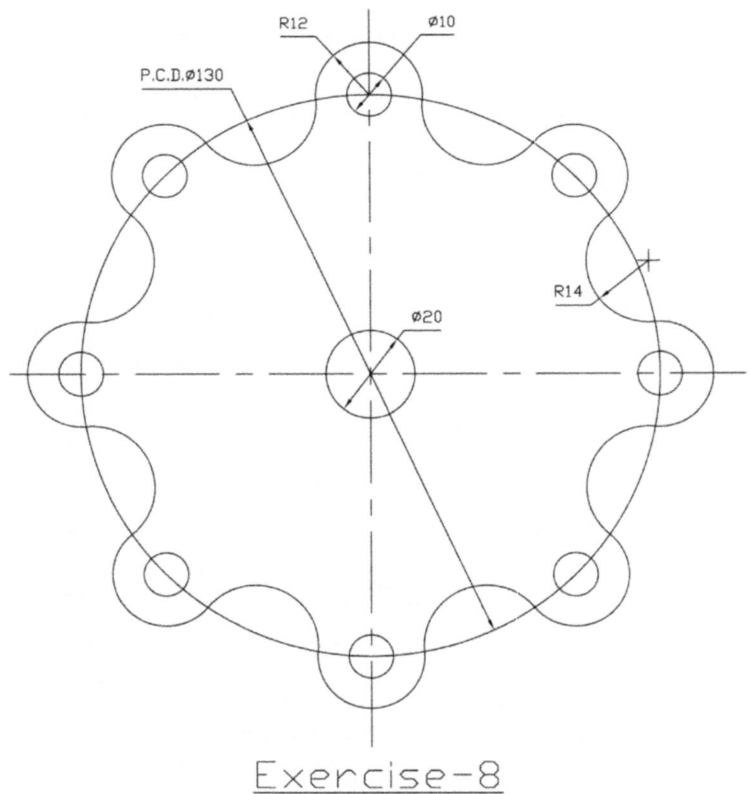

Exercise-8

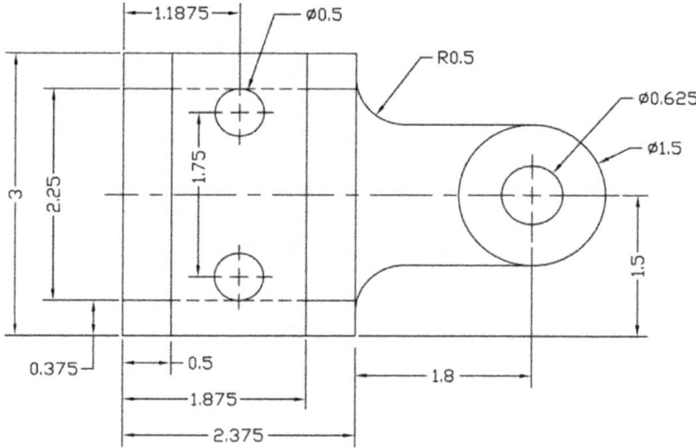

Exercise-9

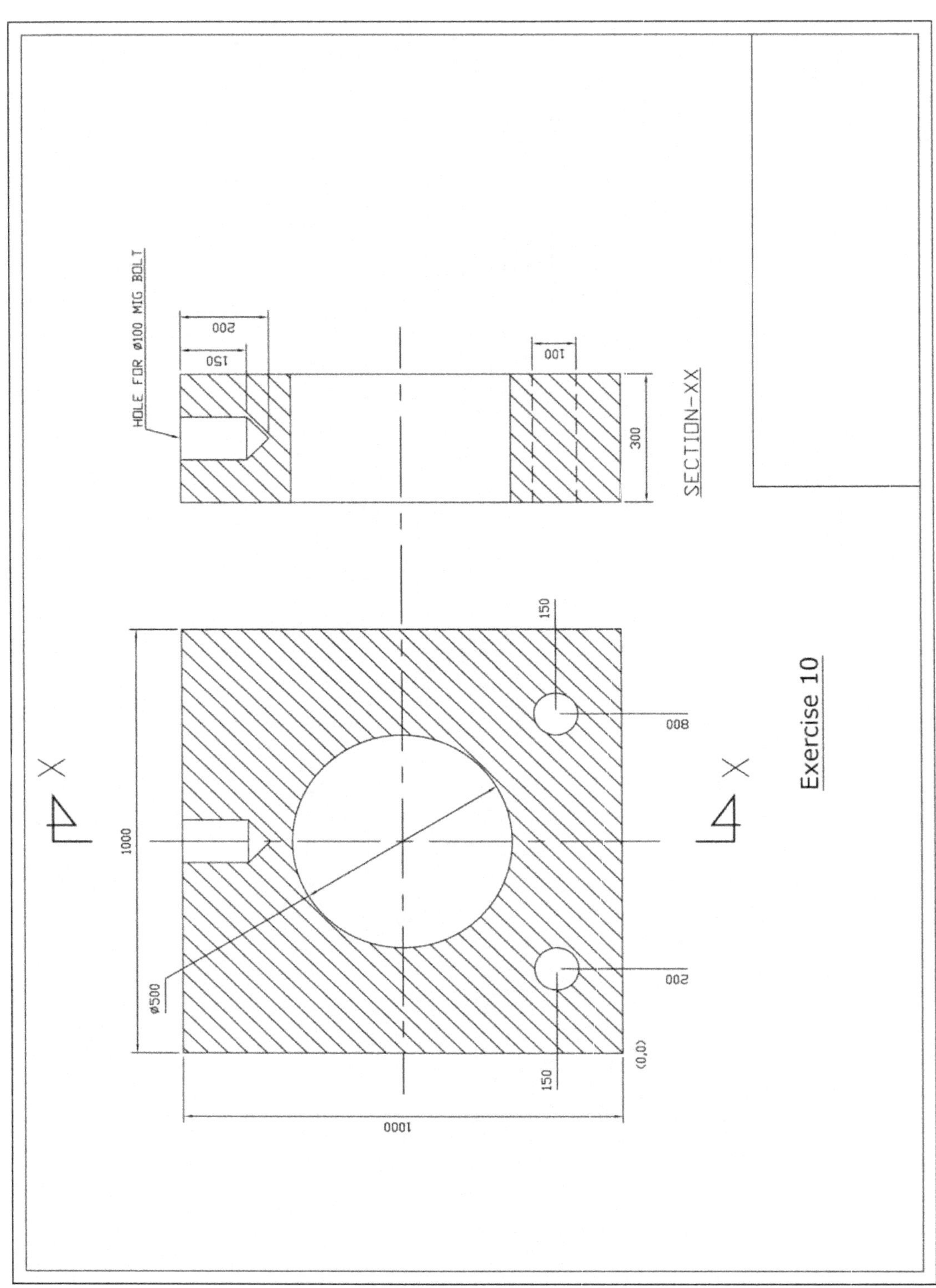
HOLE FOR Ø100 MIG BOLT
200
150
100
300
SECTION-XX
150
800
200
150
1000
Ø500
1000
(0,0)
X
X
Exercise 10

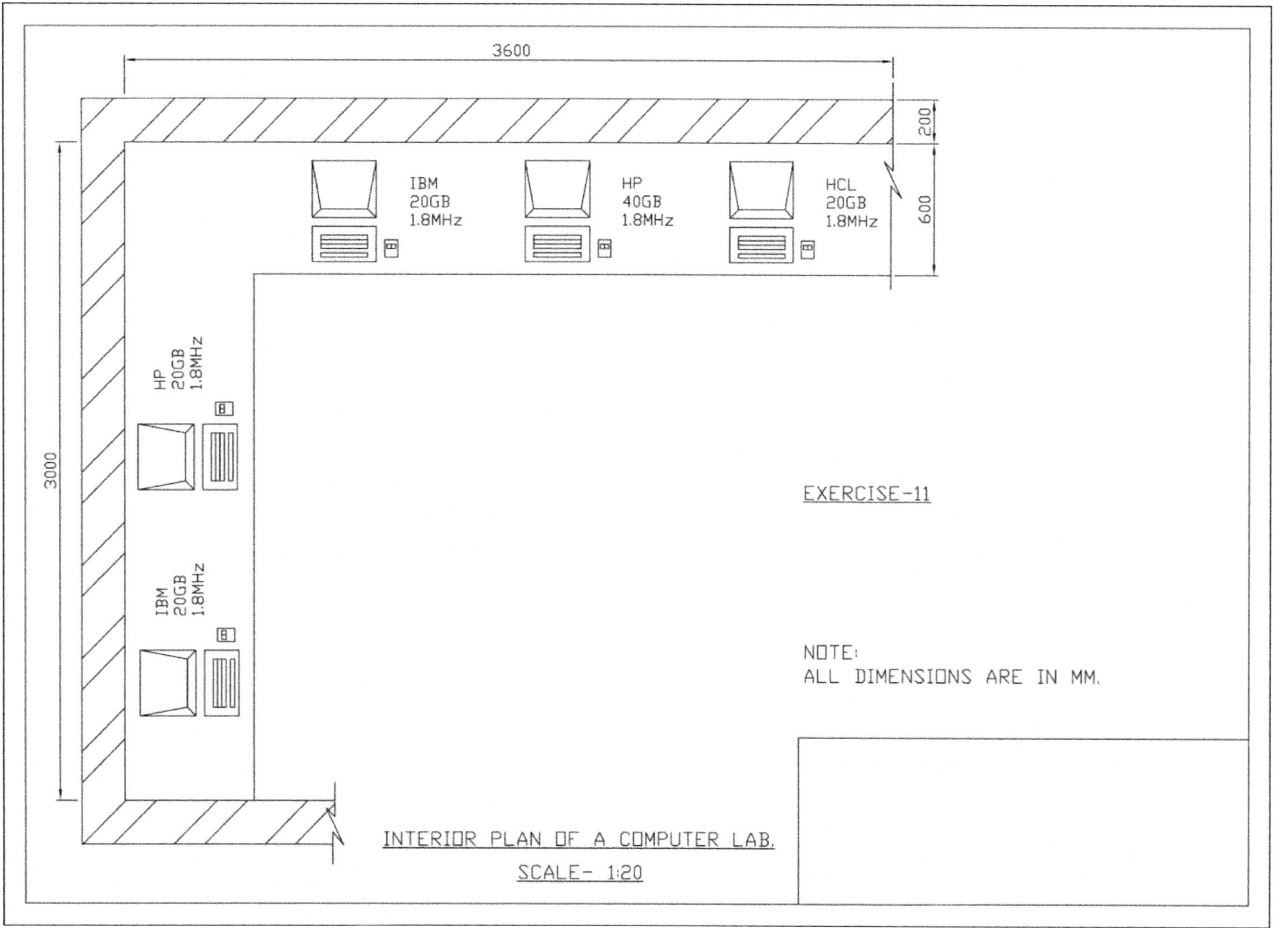

3600
200
600
IBM
20GB
1.8MHz
HP
40GB
1.8MHz
HCL
20GB
1.8MHz
3000
HP
20GB
1.8MHz
IBM
20GB
1.8MHz
EXERCISE-11
NOTE:
ALL DIMENSIONS ARE IN MM.
INTERIOR PLAN OF A COMPUTER LAB.
SCALE- 1:20

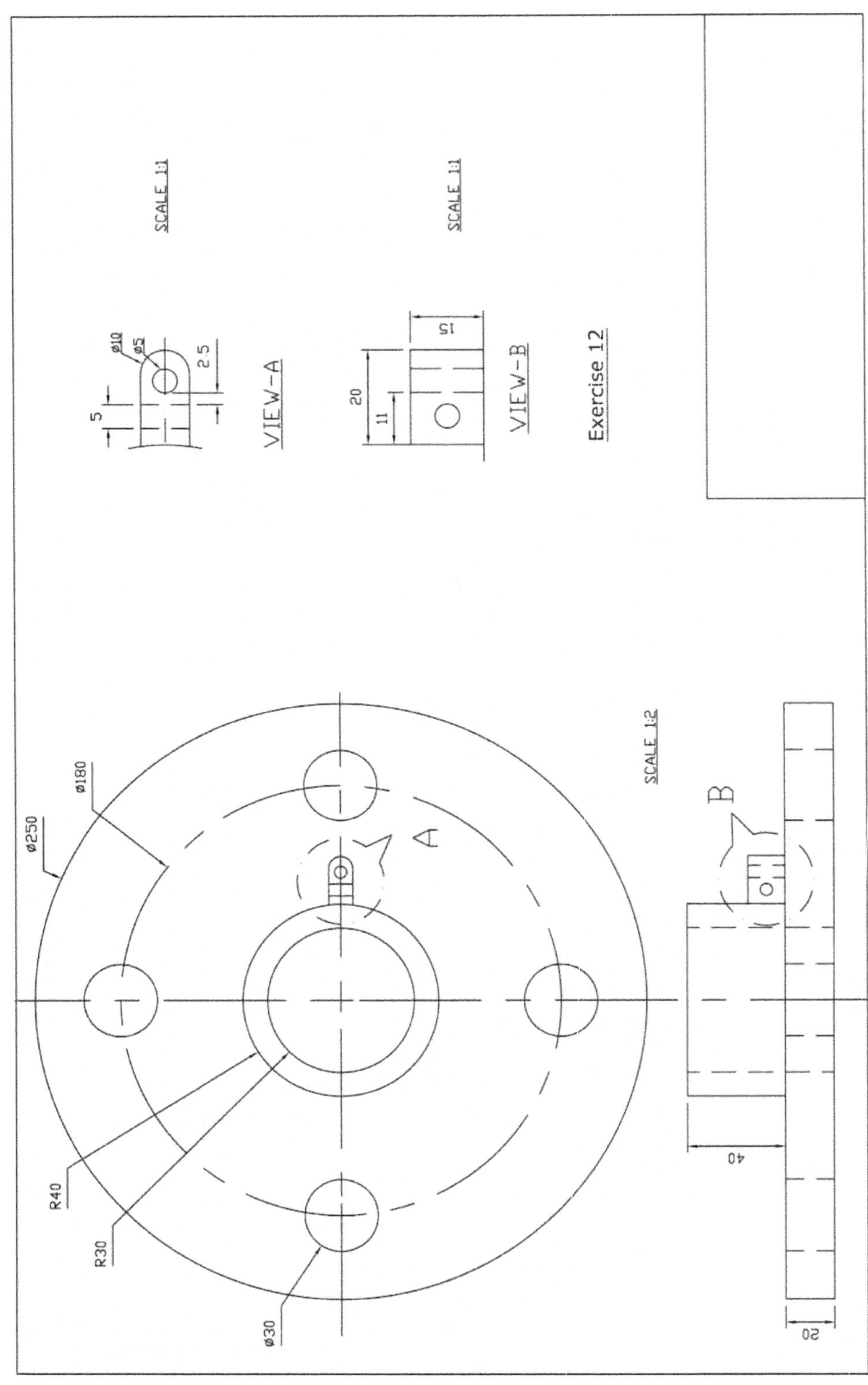
SCALE 1:1
VIEW-A
SCALE 1:1
VIEW-B
Exercise 12
SCALE 1:2
A
B
ø250
ø180
R40
R30
ø30
ø10
ø5
2.5
5
20
11
15
40
20

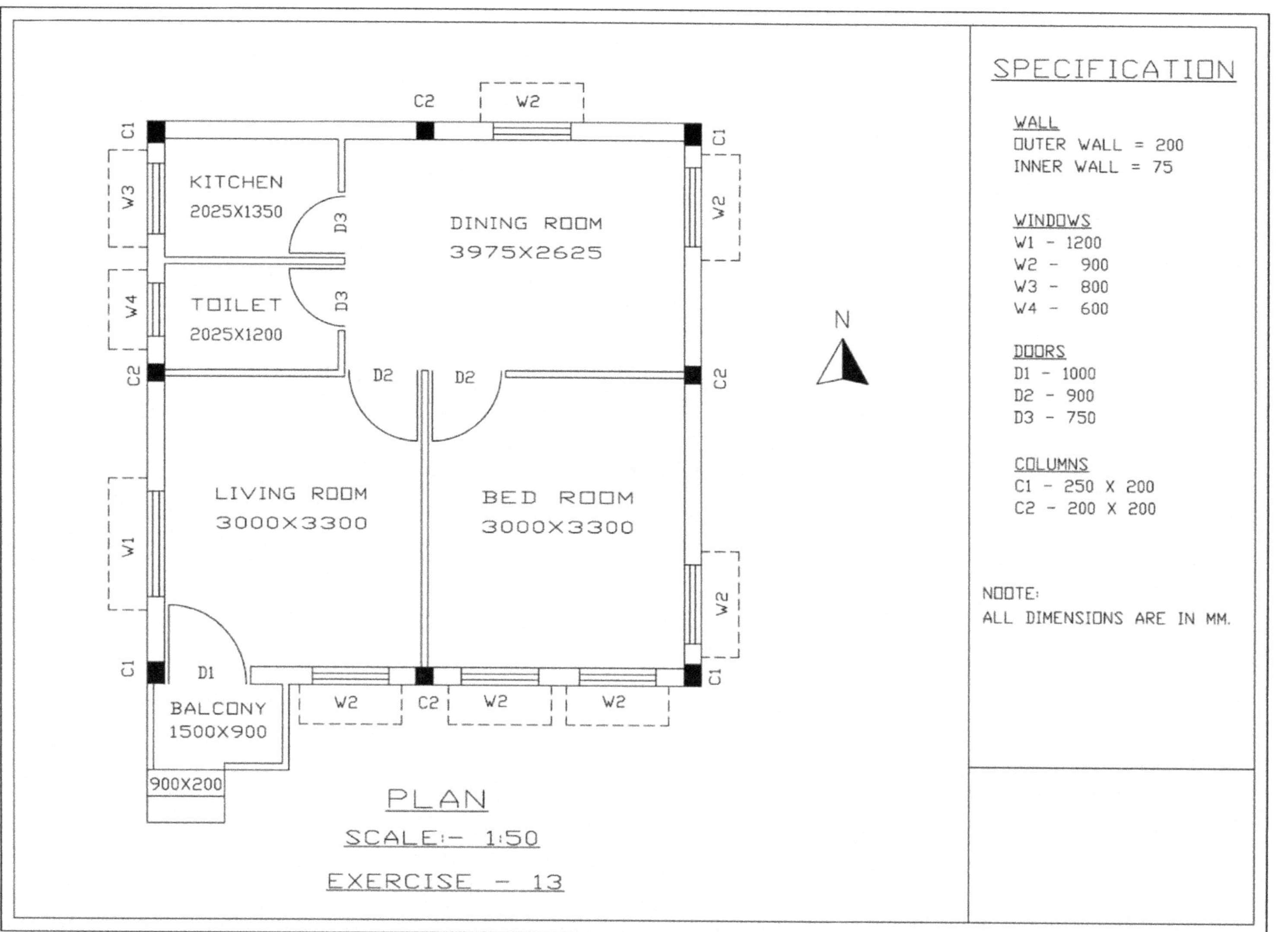
KITCHEN
2025X1350
TOILET
2025X1200
DINING ROOM
3975X2625
LIVING ROOM
3000X3300
BED ROOM
3000X3300
BALCONY
1500X900
900X200
N
PLAN
SCALE:- 1:50
EXERCISE - 13
SPECIFICATION
WALL
OUTER WALL = 200
INNER WALL = 75
WINDOWS
W1 - 1200
W2 - 900
W3 - 800
W4 - 600
DOORS
D1 - 1000
D2 - 900
D3 - 750
COLUMNS
C1 - 250 X 200
C2 - 200 X 200
NOOTE:
ALL DIMENSIONS ARE IN MM.

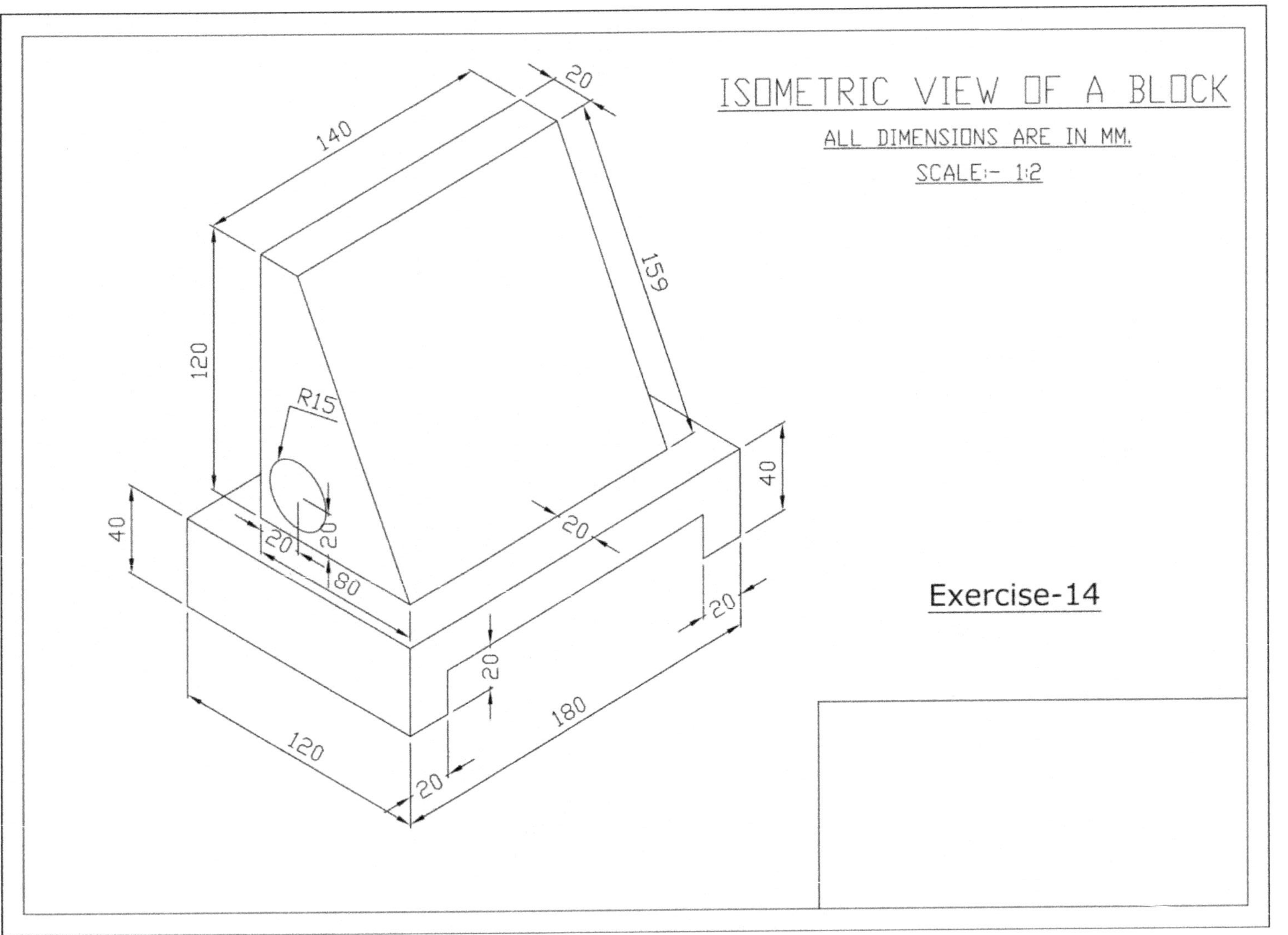
ISOMETRIC VIEW OF A BLOCK
ALL DIMENSIONS ARE IN MM.
SCALE:- 1:2
Exercise-14
140
20
159
120
R15
40
20
20
80
40
20
20
20
180
120
20

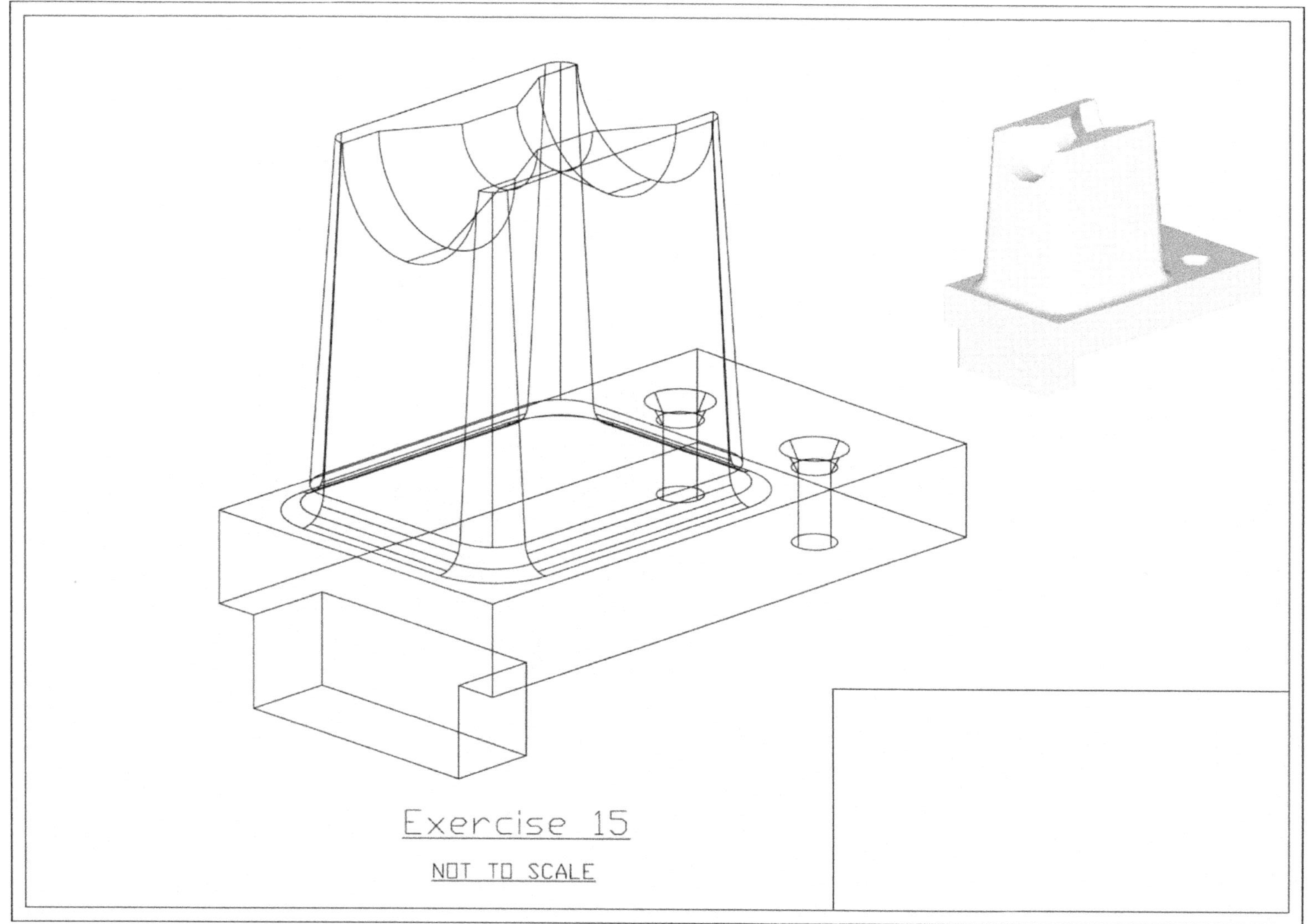
Exercise 15
NOT TO SCALE

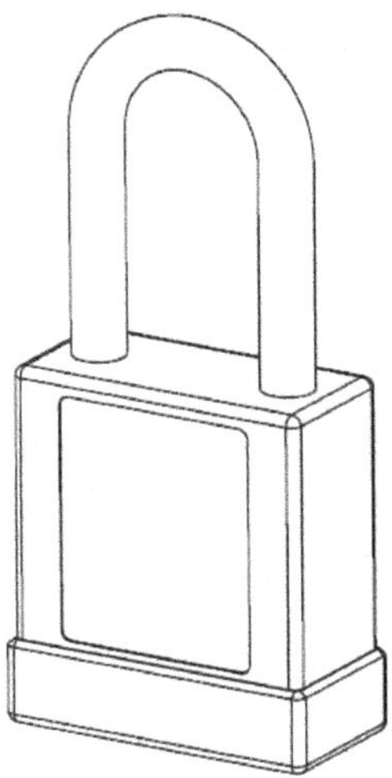

Exercise 16

Exercise 17

Index

www.ingramcontent.com/pod-product-compliance
Lightning Source LLC
Chambersburg PA
CBHW080416060326
40689CB00019B/4263

9781463625801